Implanted Antennas in Medical Wireless Communications

Implanted Antennas in Medical Wireless Communications
Yahya Rahmat-Samii and Jaehoon Kim

978-3-031-00403-2   paper   Rahmat-Samii/Kim
978-3-031-01531-1   ebook   Rahmat-Samii/Kim

DOI 10.1007/978-3-031-01531-1

A Publication in the Springer series
SYNTHESIS LECTURES ON ANTENNAS
Lecture #1
Series Editor: Constantine A. Balanis

First Edition
10 9 8 7 6 5 4 3 2 1

Implanted Antennas in Medical Wireless Communications

Yahya Rahmat-Samii and Jaehoon Kim
Department of Electrical Engineering,
University of California at Los Angeles

SYNTHESIS LECTURES ON ANTENNAS #1

iv

ABSTRACT

One of the main objectives of this lecture is to summarize the results of recent research activities of the authors on the subject of implanted antennas for medical wireless communication systems. It is anticipated that ever sophisticated medical devices will be implanted inside the human body for medical telemetry and telemedicine. To establish effective and efficient wireless links with these devices, it is pivotal to give special attention to the antenna designs that are required to be low profile, small, safe and cost effective. In this book, it is demonstrated how advanced electromagnetic numerical techniques can be utilized to design these antennas inside as realistic human body environment as possible. Also it is shown how simplified models can assist the initial designs of these antennas in an efficient manner.

KEYWORDS

Finite difference time domain, Human interaction, Implantable antenna, Medical wireless communication, Miniaturized antennas, Planar antennas, Spherical dyadic Green's function

Contents

CHAPTER 1

Implanted Antennas for Wireless Communications

1.1 INTRODUCTION

The demand to utilize radio frequency antennas inside/outside a human body has risen for biomedical applications [1–3]. Most of the research on antennas for medical applications has focused on producing hyperthermia for medical treatments and monitoring various physiological parameters [1]. Antennas used to elevate the temperature of cancer tissues are located inside or outside of the patient's body, and the shapes of antennas used depend on their locations. For instance, waveguide or low-profile antennas are externally positioned, and monopole or dipole antennas transformed from a coaxial cable are designed for internal use [1]. In addition to medical therapy and diagnosis, telecommunications are regarded as important functions for implantable medical devices (pacemakers, defibrillators, etc.) which need to transmit diagnostic information [3]. In contrast to the number of research accomplishments related to hyperthermia, work on antennas used to build the communication links between implanted devices and exterior instrument for biotelemetry are not widely reported.

It is commonly recognized that modern wireless technology will play an important role in making telemedicine possible. In not a distant future, remote health-care monitoring by wireless networks will be a feasible treatment for patients who have chronic disease (Parkinson or Alzheimer) [4]. To establish the required communication links for biomedical devices (wireless electrocardiograph, pacemaker), radio frequency antennas that are placed inside/outside of a human body need to be electromagnetically characterized through numerical and experimental techniques.

One of the main objectives of this book is to summarize the results of recent research activities of the authors on the subject of implanted antennas for medical wireless communication systems. It is anticipated that ever sophisticated medical devices will be implanted inside the human body for medical telemetry and telemedicine. To establish effective and efficient wireless links with these devices, it is pivotal to give special attention to the antenna designs that are required to be low profile, small, safe and cost effective. In this book, it is demonstrated how advanced electromagnetic numerical techniques can be utilized to design these antennas inside as realistic human body environment as possible. Also it is shown how simplified models can assist the initial designs of these antennas in an efficient manner.

1.2   CHARACTERIZATION OF IMPLANTED ANTENNAS

Figure 1.1 shows the schematic diagram of the research activities for implanted antennas inside a human body for wireless communication applications. Implanted antennas are located inside a human head and a human body and are characterized using two different numerical

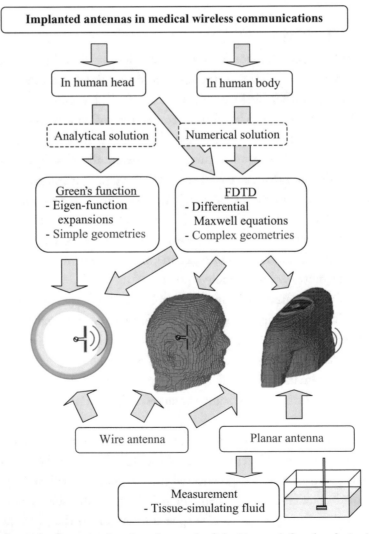

FIGURE 1.1: Schematic diagram showing the methodologies used for the designing of implanted antennas for wireless communications in this book

methodologies (spherical dyadic Green's function (DGF) and finite difference time domain (FDTD)). There are clearly other numerical techniques that can be used. If an antenna is positioned in a human head, the characteristic data for the antenna is obtained using spherical DGF expansions because the human head can be simplified as a lossy multi-layered sphere. This simplification provides useful capability to perform parametric studies. Numerical methodologies (spherical DGF and FDTD) are implemented to characterize antennas inside a human head/body and to design implanted low-profile antennas to establish medical communication links between active medical implantable devices and exterior equipment.

For medical wireless communication applications, implanted antennas operate at the medical implant communications service (MICS) frequency band (402–405 MHz) which is regulated by the Federal Communication Commission (FCC) [5] and the European Radio-communications Committee (ERC) for ultra low power active medical implants [6].

1.2.1   Antennas Inside Biological Tissues

For implantable communication links between implanted antennas inside a human body and exterior antennas in free space, implanted antennas are located in biological tissues in two ways. As shown in Fig. 1.2, one way is that an implanted antenna directly contacts a biological tissue and the other is that an antenna indirectly contacts a biological tissue using a buffer layer. The buffer layer of Fig. 1.2(b) can be an air region or a dielectric material. The antenna of Fig. 1.2(a) requires smaller space in a human body than that of Fig. 1.2(b), but the link of Fig. 1.2(a) generates higher SAR value because of the direct contact. The advantage of Fig. 1.2(b) link is

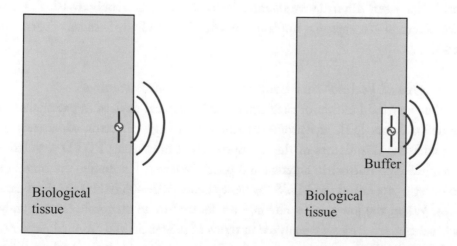

(a) Directly contacting the biological tissue.   (b) Indirectly contacting the biological tissue.

FIGURE 1.2: Two different antenna configurations inside the biological tissue

that there exist many possible methods to improve the performance of the communication link through diverse electrical characterization as it will be shown later.

1.2.2   Spherical Dyadic Green's Function

For the spherical dyadic Green's Function (DGF) simulations, a human head is approximated as a multi-layered lossy sphere with material characteristics based on measured data [7]. The expressions for the field distributions of the antenna inside the inhomogeneous sphere are obtained using the spherical DGF [8, 9]. By applying the infinitesimal current decomposition of the implanted antenna [10, 11] and introducing rotation of the coordinate system, the general expressions of the spherical DGF are modified to construct the required numerical codes. The law of energy conservation and the comparison of the results with the finite difference time domain (FDTD) simulations are used to verify the accuracy of the spherical DGF code.

1.2.3   Finite Difference Time Domain

For the FDTD analysis, the phantom data for a human body produced by computer tomography (CT) and the electric characteristic data of human biological tissues are combined to represent the input file for the computer simulations. The near-field distributions calculated from the spherical DGF code are compared with those from the FDTD code in order to evaluate the viability of the spherical DGF methodology for the analysis of implanted antennas inside a human head. To check how the human body affects the radiation characteristics of an implanted dipole in a human head, a three-dimensional geometry for the FDTD simulations was also constructed to include a human shoulder.

Beside characterization of wire antennas inside a human head or body, FDTD simulations are used to design planar antennas implanted inside a human body because of versatility of the FDTD code.

1.2.4   Design and Performance Evaluations of Planar Antennas

Based on the expected location of such implantable medical devices as pacemakers and cardioverter defibrillators [12], low-profile antennas with high dielectric superstrate layers are designed under the skin tissues of the left upper chest area using FDTD simulations. Two antennas (spiral-type microstrip antenna and planar inverted F antenna) are tuned to a 50 Ω system in order to operate at the MICS frequency band (402–405 MHz) for short-range medical devices. When the low-profile antennas are located in an anatomic human body model, their electrical characteristics are analyzed in terms of near-field and far-field patterns.

A FDTD simulation geometry simplified from an anatomic human body is utilized to facilitate the design of implanted planar inverted planar F antennas (PIFA). PIFAs are constructed using printed circuit technology and are fed by a coaxial cable. To measure impedance matching characteristics of the constructed implanted antennas to a 50 Ω system, the constructed antennas are inserted inside a tissue-simulating fluid whose electrical characteristics are very similar to those of the biological tissues. Maximum available power is calculated to analyze the reliability of the communication link and is used to estimate the minimum sensitivity requirement for receiving systems.

For the evaluation of radiation performances and safety issues related to implanted antennas, the radiation characteristics and 1-g averaged specific absorption rate (SAR) distributions are simulated and compared with ANSI/IEEE limitations for SAR [13]. Additionally, the numerical computational procedures recommended by IEEE [14] are applied to extract SAR values for implanted antennas.

CHAPTER 2

Computational Methods

In this chapter, the numerical methodologies discussed before (Green's function and finite difference time domain (FDTD)) are utilized to characterize the antennas implanted inside a human body. To analyze the antennas in a human head, spherical Green's functions are expressed in terms of spherical vector wave functions in order to obtain the closed form analytical formula. In addition, the input file for FDTD simulations which utilizes the integral form of Maxwell equations is generated.

2.1   GREEN'S FUNCTION METHODOLOGY

When an antenna is positioned in a human head, the characteristic data for the antenna is analytically obtained by simplifying the human head as a lossy multi-layered sphere. To facilitate the numerical implementation of spherical Green's function codes, infinitesimal current decomposition of the implanted source [10] and rotation of the coordinate system are utilized to modify the general spherical DGF expressions given in [9].

2.1.1   Spherical Head Models

A human head is represented as a multi-layered lossy dielectric sphere consisting of skin, fat, bone, dura, cerebrospinal fluid (CSF) and brain whose structure is shown in Fig. 2.1. Table 2.1 shows the electric characteristics (relative dielectric constant and conductivity) of the biological tissue in the model of the spherical head at 402 MHz using measured data from [8].

By changing the parameters of the spherical head models, different kinds of spherical human head models are generated to represent a human head. As shown in Table 2.2, three kinds of spherical head models are given: homogenous head [11, 15], three-layered head [11–15], six-layered head model [16]. The homogeneous head has a brain tissue layer only, the three-layered head consists of brain, bone, skin layers, and the six-layered head consists of skin, fat, bone, dura, cerebrospinal fluid (CSF) and brain layers. Each head model commonly has the brain tissue and the total size is the same.

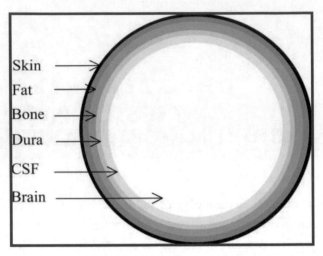

FIGURE 2.1: Schematic of the six-layered spherical head modeling a human head

TABLE 2.1: Electrical Characteristics of the Biological Tissues in the Spherical Head Models at 402 MHz

BIOLOGICAL TISSUE	PERMITTIVITY (ε_r)	CONDUCTIVITY (σ, S/m)
Brain	49.7	0.59
CSF	71.0	2.25
Dura	46.7	0.83
Skull	17.8	0.16
Fat	5.6	0.04
Skin	46.7	0.69

TABLE 2.2: Single-, Three- and Six-layered Spherical Head Models with an Outer Radius of 9 cm

BIOLOGICAL TISSUE	HOMOGENEOUS HEAD (cm)	THREE-LAYERED HEAD (cm)	SIX-LAYERED HEAD (cm)
Brain	$a_1 = 9.00$	$a_3 = 8.10$	$a_6 = 8.10$
CSF			$a_5 = 8.30$
Dura			$a_4 = 8.35$
Bone		$a_2 = 8.55$	$a_3 = 8.76$
Fat			$a_2 = 8.90$
Skin		$a_1 = 9.00$	$a_1 = 9.00$

2.1.2 Spherical Green Function's Expansion

By modeling a human head as a sphere consisting of multiple layers of different lossy dielectric materials, the antenna implanted in a human head is represented as the current source in the multi-layered sphere as shown in Fig. 2.2. The total electric field, $\overline{E}_f$ in the field layer generated by current density of the implanted source, $\overline{J}_s(\overline{r}')$ in the source layer can be calculated by the volume integration in terms of the spherical dyadic Green's Functions (DGF) [8]:

$$\overline{E}_f(\overline{r}) = -j\omega\mu_s \iiint_v \left[\overline{\overline{G}}_{e0}(\overline{r}, \overline{r}')\delta_{fs} + \overline{\overline{G}}_{es}^{(fs)}(\overline{r}, \overline{r}') \right] \cdot \overline{J}_s(\overline{r}')dv' \qquad (2.1)$$

where ω is the angular frequency, μ_s the permeability of the source layer, $\overline{r} = (r, \theta, \phi)$ the field location, $\overline{r}' = (r_0, \theta_0, \phi_0)$ the source location, and δ_{fs} the Kronecker delta. In addition, $\overline{\overline{G}}_{e0}(\overline{r}, \overline{r}')$ is an unbounded spherical DGF in the source region, and $\overline{\overline{G}}_{es}^{(fs)}(\overline{r}, \overline{r}')$ is a scattering spherical DGF for the field in the field layer from the current source in the source

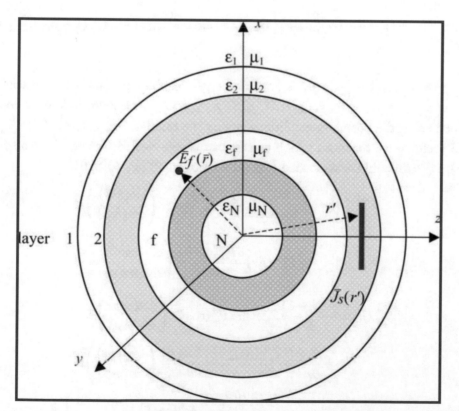

FIGURE 2.2: Multi-layered spherical head model with an arbitrarily located electrical current source

layer [9]:

$$
\overline{\overline{G}}_{e0}(\overline{r}, \overline{r}') = \frac{1}{\beta_s^2}\hat{r}\hat{r}\delta(\overline{r} - \overline{r}') - \frac{j\beta_s}{4\pi}\sum_{n=1}^{\infty}\sum_{m=0}^{n}(2 - \delta_{m0})\frac{(2n+1)(n-m)!}{n(n+1)(n+m)!}
$$
$$
\times \begin{cases} \left[\overline{M}_{e\,mn}^{(4)}(\beta_s)\overline{M}_{e\,mn}^{(1)'}(\beta_s) + \overline{N}_{e\,mn}^{(4)}(\beta_s)\overline{N}_{e\,mn}^{(1)'}(\beta_s)\right] & r > r_0 \\[2mm] \left[\overline{M}_{e\,mn}^{(1)}(\beta_s)\overline{M}_{e\,mn}^{(4)'}(\beta_s) + \overline{N}_{e\,mn}^{(1)}(\beta_s)\overline{N}_{e\,mn}^{(4)'}(\beta_s)\right] & r < r_0 \end{cases}
$$
(2.2)

$$
\overline{\overline{G}}_{es}^{(fs)}(\overline{r}, \overline{r}') = -\frac{j\beta_s}{4\pi}\sum_{n=1}^{\infty}\sum_{m=0}^{n}(2 - \delta_{m0})\frac{(2n+1)(n-m)!}{n(n+1)(n+m)!}
$$
$$
\times \left\{(1 - \delta_{fN})\overline{M}_{e\,mn}^{(4)}(\beta_f)\left[A_{n,f}(1 - \delta_{sN})\overline{M}_{e\,mn}^{(4)'}(\beta_s) + B_{n,f}(1 - \delta_{s1})\overline{M}_{e\,mn}^{(1)'}(\beta_s)\right]\right.
$$
$$
+ (1 - \delta_{fN})\overline{N}_{e\,mn}^{(4)}(\beta_f)\left[C_{n,f}(1 - \delta_{sN})\overline{N}_{e\,mn}^{(4)'}(\beta_s) + D_{n,f}(1 - \delta_{s1})\overline{N}_{e\,mn}^{(1)'}(\beta_s)\right]
$$
$$
+ (1 - \delta_{f1})\overline{M}_{e\,mn}^{(1)}(\beta_f)\left[E_{n,f}(1 - \delta_{sN})\overline{M}_{e\,mn}^{(4)'}(\beta_s) + F_{n,f}(1 - \delta_{s1})\overline{M}_{e\,mn}^{(1)'}(\beta_s)\right]
$$
$$
\left. + (1 - \delta_{f1})\overline{N}_{e\,mn}^{(1)}(\beta_f)\left[G_{n,f}(1 - \delta_{sN})\overline{N}_{e\,mn}^{(4)'}(\beta_s) + H_{n,f}(1 - \delta_{s1})\overline{N}_{e\,mn}^{(1)'}(\beta_s)\right]\right\}
$$
(2.3)

where $\beta_s = \omega\sqrt{\mu_s\epsilon_s(1 - \frac{j\sigma_s}{\omega\epsilon_s})}$, $\mu_s, \epsilon_s, \sigma_s$ are the permeability, permittivity, and conductivity of the source layer, $A_{n,f}, B_{n,f}, C_{n,f}, D_{n,f}, E_{n,f}, F_{n,f}, G_{n,f}, H_{n,f}$ are unknown coefficients determined using the boundary conditions among the multi-layers, and $\overline{M}_{e\,mn}^{(i)}, \overline{N}_{e\,mn}^{(i)}$ are the spherical vector wave functions which consist of the spherical Bessel and Hankel functions $Z_n^{(1)}(\beta r), Z_n^{(4)}(\beta r)$ and the associated Legendre function $P_n^m(\cos\theta)$[8, 9]:

$$
\overline{M}_{e\,mn}^{(i)}(\beta) = \mp\frac{m}{\sin\theta}Z_n^{(i)}(\beta r)P_n^m(\cos\theta)\begin{pmatrix} \sin(m\phi) \\ \cos(m\phi) \end{pmatrix}\hat{\theta}
$$
$$
- Z_n^{(i)}(\beta r)\frac{\partial P_n^m(\cos\theta)}{\partial\theta}\begin{pmatrix} \cos(m\phi) \\ \sin(m\phi) \end{pmatrix}\hat{\phi}
$$
(2.4)

$$
\overline{N}_{e\,mn}^{(i)}(\beta) = \frac{n(n+1)}{\beta r}Z_n^{(i)}(\beta r)P_n^m(\cos\theta)\begin{pmatrix} \cos(m\phi) \\ \sin(m\phi) \end{pmatrix}\hat{r}
$$
$$
+ \frac{1}{\beta r}\frac{\partial[rZ_n^{(i)}(\beta r)]}{\partial r}\frac{\partial P_n^m(\cos\theta)}{\partial\theta}\begin{pmatrix} \cos(m\phi) \\ \sin(m\phi) \end{pmatrix}\hat{\theta}
$$
(2.5)
$$
\mp \frac{m}{\beta r\sin\theta}\frac{\partial[rZ_n^{(i)}(\beta r)]}{\partial r}P_n^m(\cos\theta)\begin{pmatrix} \sin(m\phi) \\ \cos(m\phi) \end{pmatrix}\hat{\phi}
$$

2.1.3 Simplification of Spherical Green's Function Expansion

Two techniques are utilized to simplify the calculation of the Green's function expansion in the form of Eq. (2.1). The first technique in reducing the complexity of the volume integration is to represent the dipole antenna as superposition of infinitesimal current elements lining up along the antenna [10, 11], as shown in Fig. 2.3. Furthermore, by rotating the coordinate system, one is able to decompose each current element into its local x_1-directed and z_1-directed components on the assumption that the dipole is positioned in the x–z plane [17]. In Fig. 2.3, $Il_x(r'_l, 0, 0)$ and $Il_z(r'_l, 0, 0)$ represent the x-directed and z-directed current moments decomposed from the original current moment using the rotation of the coordinate system.

Based on the decomposition of the dipole antenna and the rotation of the coordinate system for each current element, the electric field expression based on the local (rotated) coordinate system (x_l, y_l, z_l) from each infinitesimal current moment modified from Eq. (2.1) is given as:

$$\overline{E}_f(\overline{r}_l) = -j\omega\mu_f \left\{ \overline{\overline{G}}_{e0}(\overline{r}_l, \overline{r}'_l)\delta_{fs} + \overline{\overline{G}}_{es}^{(fs)}(\overline{r}_l, \overline{r}'_l) \right\} \cdot \{\hat{x}_l Il_x + \hat{z}_l Il_z\}$$

$$= \overline{E}_x^i(\overline{r}_l) + \overline{E}_x^s(\overline{r}_l) + \overline{E}_z^i(\overline{r}_l) + \overline{E}_z^s(\overline{r}_l) \qquad (2.6)$$

where $\overline{r}_l = (r_l, \theta_l, \phi_l)$, $\overline{r}'_l = (r'_l, \theta'_l, \phi'_l)$ and $\overline{E}_x^i(\overline{r}_l)$ and $\overline{E}_x^s(\overline{r}_l)$ are the incident and scattering electric fields from an x-directed infinitesimal electric current moment on the z-axis. Similarly, $\overline{E}_z^i(\overline{r}_l)$ and $\overline{E}_z^s(\overline{r}_l)$ represent the incident and scattering electric fields from a z-directed infinitesimal electric current moment on the z-axis.

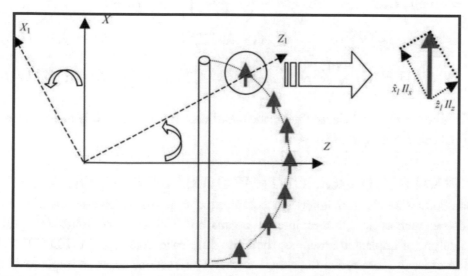

FIGURE 2.3: Decomposition of the finite length dipole and rotation of the coordinate system for each infinitesimal current element

Because x direction ($\hat{x}_l$) on the z-axis is equal to the θ direction ($\hat{\theta}_l$) in the local spherical coordinate system, the scattering electric field expression from an x-directed infinitesimal electric current moment on the z-axis is reformulated as:

$$\bar{E}_x^s(\bar{r}_l) = -j\omega\mu_f Il_x \overline{\overline{G}}_{es}^{(fs)}(\bar{r}_l, \bar{r}_l') \cdot \hat{\theta}_l = -\left(\frac{\omega\mu_f Il_x}{4\pi}\right)\frac{1}{r_l'}\sum_{n=1}^{\infty}\frac{(2n+1)}{n(n+1)}$$

$$\times\begin{bmatrix} (1-\delta_{fN})\overline{M}_{o1n}^{(4)}(\beta_f)\left[A_{n,f}(1-\delta_{sN})\rho_{r_l'}Z_{(n)}^{(4)}(\rho_{r_l'}) + B_{n,f}(1-\delta_{s1})\rho_{r_l'}Z_{(n)}^{(1)}(\rho_{r_l'})\right] \\ +(1-\delta_{fN})\overline{N}_{e1n}^{(4)}(\beta_f)\left[C_{n,f}(1-\delta_{sN})\frac{\partial[\rho_{r_l'}Z_n^{(4)}(\rho_{r_l'})]}{\partial(\rho_{r_l'})} + D_{n,f}(1-\delta_{s1})\frac{\partial[\rho_{r_l'}Z_n^{(1)}(\rho_{r_l'})]}{\partial(\rho_{r_l'})}\right] \\ +(1-\delta_{f1})\overline{M}_{o1n}^{(1)}(\beta_f)\left[E_{n,f}(1-\delta_{sN})\rho_{r_l'}Z_{(n)}^{(4)}(\rho_{r_l'}) + F_{n,f}(1-\delta_{s1})\rho_{r_l'}Z_{(n)}^{(1)}(\rho_{r_l'})\right] \\ +(1-\delta_{f1})\overline{N}_{e1n}^{(1)}(\beta_f)\left[G_{n,f}(1-\delta_{sN})\frac{\partial[\rho_{r_l'}Z_n^{(4)}(\rho_{r_l'})]}{\partial(\rho_{r_l'})} + H_{n,f}(1-\delta_{s1})\frac{\partial[\rho_{r_l'}Z_n^{(1)}(\rho_{r_l'})]}{\partial(\rho_{r_l'})}\right] \end{bmatrix}$$

$$(2.7)$$

where $\rho_{r_l'} = \beta_s r_l'$.

Similarly, because z direction ($\hat{z}_l$) on the z-axis is equal to the r direction ($\hat{r}_l$) in the local spherical coordinate system, the scattering electric field from a z-directed infinitesimal electric current on the z-axis is given as:

$$\bar{E}_z^s(\bar{r}_l) = -j\omega\mu_f Il_z \overline{\overline{G}}_{es}^{(fs)}(\bar{r}_l, \bar{r}_l') \cdot \hat{r}_l = -\left(\frac{\omega\mu_f Il_z}{4\pi}\right)\frac{1}{r_l'}\sum_{n=1}^{\infty}(2n+1)$$

$$\times\begin{bmatrix}(1-\delta_{fN})\overline{N}_{e0n}^{(4)}(\beta_f) & \left[C_{n,f}(1-\delta_{sN})Z_n^{(4)}(\rho_{r_l'}) + D_{n,f}(1-\delta_{s1})Z_n^{(1)}(\rho_{r_l'})\right] \\ +(1-\delta_{f1})\overline{N}_{e0n}^{(1)}(\beta_f)\left[G_{n,p}(1-\delta_{sN})Z_n^{(4)}(\rho_{r_l'}) + H_{n,f}(1-\delta_{s1})Z_n^{(1)}(\rho_{r_l'})\right]\end{bmatrix}\quad(2.8)$$

To restore the actual electric field, the coordinate transformation is needed to return to the original coordinate system (x, y, z).

2.2   FINITE DIFFERENCE TIME DOMAIN METHODOLOGY

A human body is an electromagnetically complicated structure which consists of various biological tissues such as skins, bones, internal organs, etc. To include complex biological tissues for the analysis of implanted antennas, the finite difference time domain (FDTD) method is utilized in order to characterize the electromagnetic interactions between implanted antennas and a human head/body and to design low-profile antennas which are able to operate in the complex environment of human body.

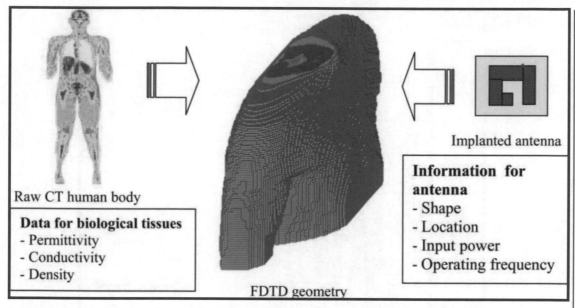

Raw CT human body

Data for biological tissues
- Permittivity
- Conductivity
- Density

FDTD geometry

Implanted antenna

Information for antenna
- Shape
- Location
- Input power
- Operating frequency

FIGURE 2.4: Schematic diagram for the FDTD input file (geometry) generation

2.2.1   Input File for FDTD Simulation

To simulate implanted antennas in a human body, the input file for FDTD codes needed to be prepared. The first step is to make an anatomical body model which is read by a FDTD computer code and the second is to locate implanted antennas inside the body model. Figure 2.4 shows how to translate the phantom data for a human body produced by raw computer tomography (CT) into the input file needed for the FDTD simulations. By using the tissue information, the proper electric characteristic data such as permittivity, conductivity, and mass density are assigned to each voxel. Antennas are properly located and operated inside a human head/body by applying specific information (the shape, location, input values) about implanted antennas. The electric and magnetic fields at every unit cell are updated by using the integral form of Maxwell equations.

2.2.2   Human Body Model

The FDTD human model in Fig. 2.5 is represented by relative permittivity, ranging from 0 to 70 at 402 MHz. For this model, the 67 biological tissue phantom file for a human body produced from computer tomography (CT) in the Yale University School of Medicine [18] was translated into the 30 biological tissue FDTD model using the available measured electrical data of biological tissues [7] given in Table 2.3. The phantom data consists of $155 \times 72 \times 487$ volume pixel (voxels) which contains the information on the biological tissues. Because the distance between neighboring voxels is 4 mm, the cell size of the FDTD model is the same as

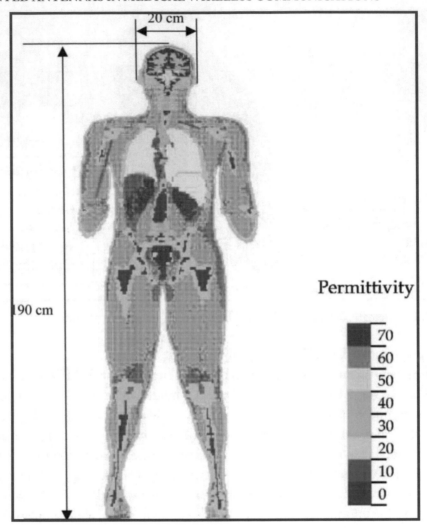

FIGURE 2.5: A human body model represented by different relative permittivity

the voxel size (4 mm) of the phantom file. According to Table 2.3, the range of the conductivity for biological tissues is from about 0 to 3 S/m, and the relative permeability for biological tissues is 1. The mass densities of biological tissues taken from [19] are between 0 and 2 g/cm^3.

2.3   NUMERICAL TECHNIQUES VERIFICATIONS BY COMPARISONS

The validations of the spherical DGF implementation for implanted antennas in a human head were accomplished using a closed form electric field equation and the finite difference time domain (FDTD) code.

TABLE 2.3: Electrical Data of Biological Tissues Used for the Human Body Model at 402 MHz

BIOLOGICAL TISSUE	PERMITTIVITY (ε_r)	CONDUCTIVITY (σ, S/m)	MASS DENSITY (g/cm^3)
Brain	49.7	0.59	1.04
Cerebrospinal fluid	71.0	2.25	1.01
Dura	46.7	0.83	1.01
Bone	13.1	0.09	1.81
Fat	11.6	0.08	0.92
Skin	46.7	0.69	1.01
Skull	17.8	0.16	1.81
Spinal cord	35.4	0.45	1.04
Muscle	58.8	0.84	1.04
Blood	64.2	1.35	1.06
Bone marrow	5.67	0.03	1.06
Trachea	44.2	0.64	1.10
Cartilage	45.4	0.59	1.10
Jaw bone	22.4	0.23	1.85
Cerebellum	55.9	1.03	1.05
Tongue	57.7	0.77	1.05
Mouth cavity	1.0	0.00	0.00
Eye tissue	57.7	1.00	1.17
Lens	48.1	0.67	1.05
Teeth	22.4	0.23	1.85
Lungs	54.6	0.68	1.05
Heart	66.0	0.97	1.05
Liver	51.2	0.65	1.05
Kidney	66.4	1.10	1.05
Stomach	67.5	1.00	1.05
Colon	66.1	1.90	1.05
Thyroid	61.5	0.88	1.05
Trachea	44.2	0.64	1.10
Spleen	63.2	1.03	1.05
Bladder	19.8	0.33	1.05

2.3.1   Comparison with the Closed Form Equation

Two electric field intensities are compared in Fig. 2.6. One is for an infinitesimal dipole placed in the free space and the other is for an infinitesimal dipole located at the center of a homogeneous lossless dielectric ($\varepsilon_r = 49.0$, $\sigma = 0.$ S/m) sphere whose radius is 9 cm. All dipoles are assumed to deliver 1 W (watt) at 402 MHz. When the dipole is located in the free space, the electric field intensity along the z-axis is calculated by the following closed form equation [20]:

$$|E(z)| = \left| jn\frac{kIl}{4\pi z}\left(1 + \frac{1}{jkz} - \frac{1}{(kz)^2}\right)\right| \tag{2.9}$$

where η is the wave impedance ($\cong 120\pi$) in the free space, k the wave number in the free space, and Il the infinitesimal current moment which is determined by the following radiated power equation:

$$P_{rad} = \eta\left(\frac{\pi}{3}\right)\left|\frac{Il}{\lambda}\right|^2 \tag{2.10}$$

When the dipole is in the dielectric sphere, the electric field intensity along the z-axis is obtained by the spherical DGF code. The dipole placed inside the dielectric sphere produces a

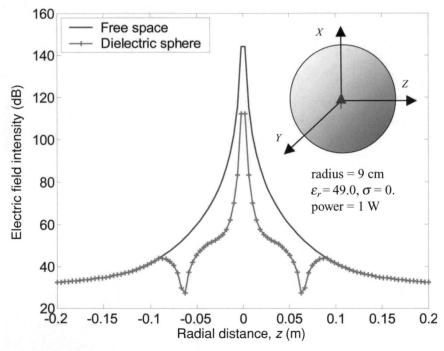

FIGURE 2.6: Comparison for the electric field intensities of infinitesimal dipoles located in the free space and in a lossless dielectric sphere

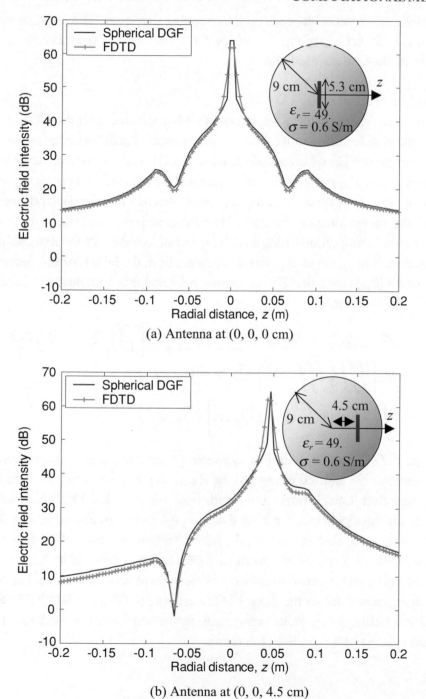

(a) Antenna at (0, 0, 0 cm)

(b) Antenna at (0, 0, 4.5 cm)

FIGURE 2.7: Comparison of the electric field intensity obtained by the spherical DGF expansions with that obtained by the FDTD code for the dipole in a homogeneous lossy sphere

standing wave pattern which depends on the operating frequency. The important observation is that both electric field distributions outside the sphere are the same because power is not dissipated in the lossless environment.

2.3.2   Comparison with FDTD

In this section, the spherical DGF implementation is compared with the FDTD simulations using the same simulation structure. For this comparison, dipole antennas are normalized to deliver the same power. The spherical code uses Eq. (2.11) in order to control the delivered power. The delivered power, P_{del} at the source point is divided into the incident power, P_{inc} delivered by the initial current and the scattered power, P_{sca} generated by the interaction between the initial current and the surrounding environment. The incident power and the scattered power are expressed by volume integrations using an unbounded spherical DGF and a scattering spherical DGF expression. Finally, the total power is generated from the initial current density, $\overline{J}_s$ of the dipole. Equation (2.9) shows that the total power delivered from antennas can be controlled by revising the initial current, $\overline{J}_s$:

$$P_{del}(\overline{r} = \overline{r}') = P_{inc}(\overline{r} = \overline{r}') + P_{sca}(\overline{r} = \overline{r}') = -\frac{1}{2}\mathrm{Re}\left[\iiint_v (\overline{E}_{inc} + \overline{E}_{sca}) \cdot \overline{J}_s^* \, dv'\right]$$

$$= -\frac{1}{2}\mathrm{Re}\left\{\iiint \left[\iiint_v -j\omega\mu_s \overline{\overline{G}}_{e0}(\overline{r}', \overline{r}')\delta_{fs} \cdot \overline{J}_s dv' \right.\right.$$

$$\left.\left. + \iiint_v -j\omega\mu_s \overline{\overline{G}}_{es}^{(fs)}(\overline{r}', \overline{r}') \cdot \overline{J}_s dv' \right] \cdot \overline{J}_s^* \, dv'\right\} \tag{2.11}$$

For the FDTD code to control the delivered power, the fact that real delivered power is equal to the sum of the radiated power and the absorbed power in a lossy medium is applied.

The near electric field distributions both from the spherical DGF code and from the FDTD code for the dipole inside a lossy dielectric sphere are compared in Fig. 2.7. For the comparison, a $0.5\lambda_d$ (dielectric wavelength, 5.3 cm) dipole was located in the homogeneous sphere whose radius is 9 cm, relative permittivity 49, relative permeability 1, and conductivity 0.6 S/m. The dipole was assumed to deliver 1 W. In spite of small differences between the two near-field distributions due to the finite FDTD cell size ($0.005\lambda_0$ at 402 MHz = 3.7 mm), both numerical techniques generate a remarkable agreement for the dipole located both at the center of the sphere and 4.5 cm from the center.

CHAPTER 3

Antennas Inside a Biological Tissue

3.1 SIMPLE WIRE ANTENNAS IN FREE SPACE

Simple wire antennas, dipoles and loops, in the free-space region are studied to examine near-field behaviors around the antennas before implanting them in biological tissues. The near-field distributions from the simple antennas in the free space are calculated in three ways: the theoretical expressions, finite difference time domain (FDTD) code, and method of moments (MoM) code to confirm the FDTD simulations which are applied to characterize the simple wire antennas inside a biological tissue.

3.1.1 Characterization of Dipole Antennas

Figure 3.1 shows a small dipole antenna located in the free space. The dipole antenna is 0.03 wavelength (λ) at 402 MHz in length and is oriented along the z-axis. The center of the coordinate system is located at the feeding point of the dipole antenna.

The electric and magnetic field magnitudes along the y-axis for the small dipole in Fig. 3.1 are theoretically expressed by Eqs. (3.1) and (3.2), respectively, which are valid field equations for 0.02–$0.1\lambda_0$ dipole antennas [20]:

$$|E(y)| = \left| \frac{I_0 l \eta}{2\pi} \left(\frac{j2\pi}{\lambda y} + \frac{1}{y^2} + \frac{\lambda}{j2\pi y^3} \right) \right| \tag{3.1}$$

$$|H(y)| = \left| \frac{I_0 l}{2\pi} \left(\frac{j2\pi}{\lambda y} + \frac{1}{y^2} \right) \right| \tag{3.2}$$

where I_0 is the maximum current of the small dipole, l the dipole's length, η the wave impedance ($= 120\pi$) in the free space and λ the wavelength. The maximum current is given in Eq. (3.3) and the radiation impedance, R_r, of the dipole is calculated in Eq. (3.4) [20]:

$$I_0 = \sqrt{\frac{2 P_{rad}}{R_r}} \tag{3.3}$$

$$R_r = 20\pi^2 \left(\frac{1}{\lambda} \right)^2 \tag{3.4}$$

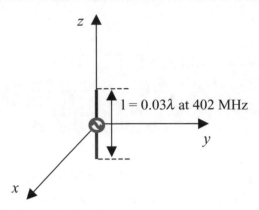

FIGURE 3.1: Small dipole antenna in the free space

According to Eqs. (3.1) and (3.2), the near electric field around the dipole antenna is proportional to the inverse cube of the radial distance while the near magnetic field is proportional to the inverse square of the radial distance. It means that electric fields would be more advantageous than magnetic field when one couples the energy from electric field sources in the near-field region.

To examine the exact field value from the small dipole, the near-field distributions of the small dipole along y-axis are calculated in three ways: the theoretical expressions, finite difference time domain (FDTD) code, and method of moments (MoM) code. In Fig. 3.2, the electric field distributions from the dipole antenna are compared. The dipole radiates 1 W into the free space and its operating frequency is 402 MHz. Three total electric field distributions in Fig. 3.2(a) are very similar except around the dipole's location which is known as a singular point. The theory generates higher electric field intensity than the real value around the singular point. By using the MoM code, Fig. 3.2(b) gives three electric field components decomposed from the total electric field. The E_x component is negligible along the y-axis. The magnitude of the E_y component only near the dipole antenna is similar to that of the E_z because circular electric field lines are generated between two polarities of the dipole. However, as the observation point extends far from the center of the dipole, the radial electric (E_y) component diminishes faster than the vertical electric field (E_z) component.

Similarly to Fig. 3.2, the magnetic fields from the small dipole antenna which radiates 1 W at 402 MHz are calculated along the y-axis in Fig. 3.3. As shown in Fig. 3.3(a), the total magnetic field obtained from the theory is well matched to those from the FDTD and MoM codes except near the singular point. Particularly, the FDTD code generates some differences around the antenna because the code uses a finite cell size. By using the MoM code, total magnetic field from the small dipole antenna is decomposed into the three magnetic field components, H_x, H_y, and H_z, as shown in Fig. 3.3(b). The magnitude of the x component is

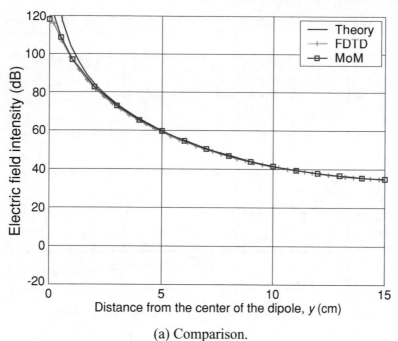

(a) Comparison.

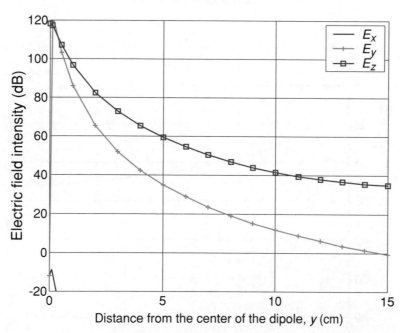

(b) Decomposition of total electric field using MoM.

FIGURE 3.2: Electric field components from the small dipole antenna in the free space

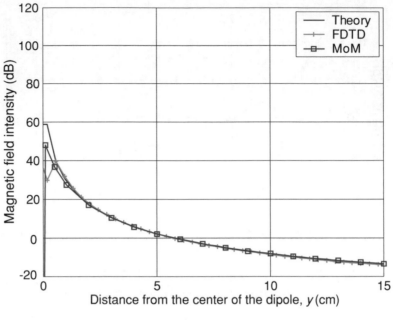

(a) Total magnetic field.

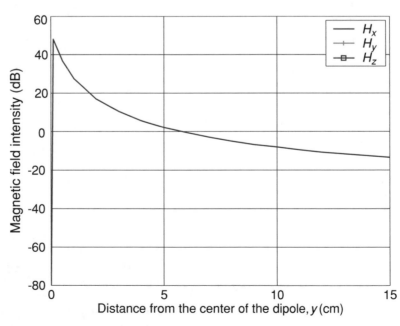

(b) Decomposition of total magnetic field using MoM.

FIGURE 3.3: Magnetic field distributions from the small dipole antenna in the free space

almost the same as that of the total magnetic field and the other components are too small to be shown in Fig. 3.3(b). Therefore, only the horizontal electric field component exists along the y-axis.

From Figs. 3.2 and 3.3, the wave impedance can be obtained by dividing the total electric field by the total magnetic field component. At 5 cm from the center of the dipole, because the total electric field intensity is about 60 dB and the total magnetic field is about 2 dB, the wave impedance is 58 dB. If the wave impedance is calculated at nearer than 5 cm, the value is higher than 58 dB. Therefore, it is expected that the wave impedance near the small dipole is much higher than the intrinsic impedance ($120\pi = 51.5$ dB) of a transverse electromagnetic (TEM) wave. It is observed that at 15 cm the wave impedance is similar to the intrinsic impedance of a TEM wave.

3.1.2   Characterization of Loop Antennas

Figure 3.4 shows a square loop antenna located in the free space. The square loop's side-width (w) is 0.03 wavelength (λ_0) at 402 MHz and the total length (l) is $0.12\lambda_0$. The origin of the coordinate system is located at the center of the loop antenna and the antenna is parallel to the x–z plane. The square loop is fed at the side of the loop, as shown in Fig. 3.4.

The magnetic field magnitude along the y-axis of the loop antenna in Fig. 3.4 is obtained from Eq. (3.5) which is a valid theoretical expression for a small circular loop antenna [20]. Because the theoretical expression for a small circular loop antenna creates a null electric field magnitude along the y-axis, the expression for the electric field magnitude is omitted:

$$|H(y)| = \left| \frac{I_0 S^2}{2\pi} \left(\frac{j2\pi}{\lambda r^2} + \frac{1}{r^3} \right) \right| \qquad (3.5)$$

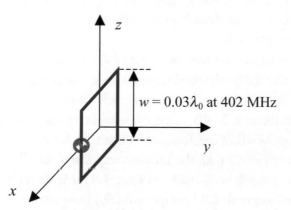

FIGURE 3.4: Square loop antenna in the free space

where I_0 is the constant current of the small loop, S the loop's area, and λ the wavelength. The constant current is given in Eq. (3.3) and the radiation impedance, R_r, of the loop is calculated in Eq. (3.6):

$$R_r = 31,171 \left(\frac{S^2}{\lambda^4} \right)^2 \tag{3.6}$$

According to Eq. (3.5), the near magnetic field around a small loop antenna varies as the inverse cube of the radial distance, similarly to a small dipole antenna whose electric field varies as the inverse cube of the radial distance.

The near electric field distributions from the square loop antennas are calculated in Fig. 3.5. The loop antenna is fed to radiate 1 W into the free space at 402 MHz. Because of the theoretical null electric field along the y-axis, the theoretical calculation is not included in Fig. 3.5. The total electric field distributions along the y-axis are calculated by the FDTD and the method of moments (MOM) codes. It is found that two simulated electric field distributions are very similar to each other. From the MoM code, total electric field from the loop antenna is decomposed into the three electric field components (E_x, E_y, and E_z), as shown in Fig. 3.5(b). As expected, the z electric component (E_z) along the y-axis is dominant and almost the same as the total magnetic field.

The near magnetic field distributions for the square loop obtained from the FDTD and MoM simulations are compared with those for a small circular loop from the theoretical expressions in Fig. 3.6. In Fig. 3.6(a), the total magnetic field from FDTD is well matched to the field from MOM although the theory generates higher field values near the antenna because the theoretical expression is for a small circular loop antenna. Therefore, it is expected that smaller loop antennas generate higher near magnetic fields. The decomposed magnetic fields are shown in Fig. 3.6(b). Because the longitudinal magnetic component, H_y along the y-axis is dominant, a receiving antenna should be properly located to maximize the power coupling from a transmitting loop antenna.

From Figs. 3.5 and 3.6, the wave impedance from the loop antenna can be obtained by dividing the total electric field by the total magnetic field component. At 5 cm from the center of the dipole, because the ratio of the total electric field to the total magnetic field is about 47 to 13 dB, the wave impedance is 34 dB. If the distance decreases to the loop antenna, the wave impedance is lower than 34 dB. Therefore, as the distance decreases to the antenna, the wave impedance becomes much lower than the intrinsic impedance of a TEM wave. It is observed that at 15 cm the wave impedance is still less than the intrinsic impedance of a TEM wave because the longitudinal magnetic field component (H_z) from the loop is much higher than the transverse components.

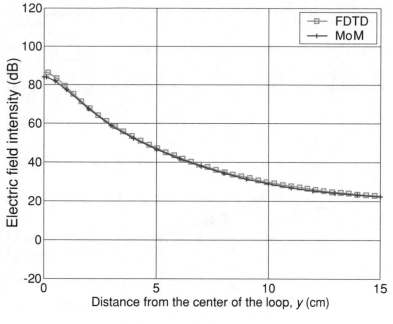

(a) Total electric field.

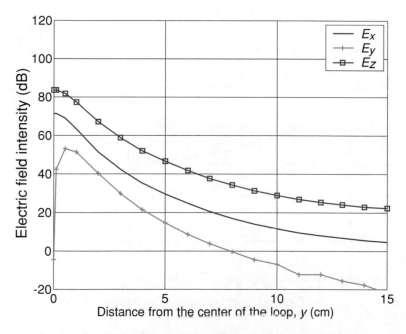

(b) Decomposition of total electric field using MoM.

FIGURE 3.5: Electric field distributions from the square loop antenna in the free space

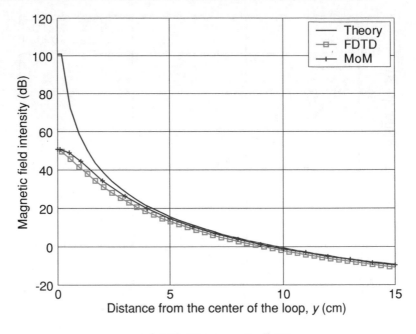

(a) Total magnetic field.

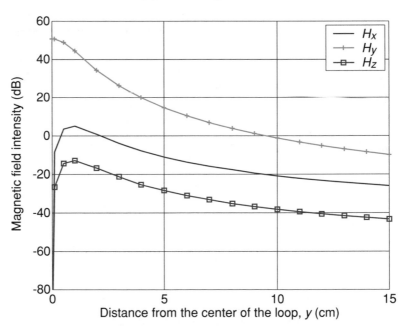

(b) Decomposition of total magnetic field using MoM.

FIGURE 3.6: Magnetic field distributions from the square loop antenna in the free space

3.2   WIRE ANTENNAS IN BIOLOGICAL TISSUE

To characterize simple wire antennas inside a biological tissue by using FDTD simulations, the dipole antenna in Fig. 3.1 and the loop antenna in Fig. 3.4 are located in a simplified biological tissue whose dimensions are 13.4 cm × 7.8 cm × 13.4 cm, as shown in Fig. 3.7. The length (0.03λ at 402 MHz = 2.2 cm) of the dipole is the same as the side-width of the loop. The simplified body model is uniformly filled with a single biological tissue whose relative permittivity (ε_r) is 49, relative permeability (μ_r) 1, and conductivity (σ) 0.6 S/m.

The antennas implanted in the simplified model is centered at an air-box whose dimensions are 3.7 cm × 0.7 cm × 3.7 cm. Because it is assumed that the implanted antennas are located under a skin biological tissue in a human body, the depth (3.7 mm) of the air-box from the free space represents the thickness of the skin tissue. The center of the air-box is the same as the centers of the wire antennas.

Figure 3.8 shows the electric and magnetic field distributions along the y-axis from the dipole antenna in the simplified tissue model. The antenna is assumed to deliver 1 W and operate at 402 MHz. At the boundary between the tissue and the free space, it is observed that the slope of the electric field is abruptly changed.

Table 3.1 shows the electric field and magnetic field variations along the radial direction (y-axis) from the dipole in the free space and the biological tissue. The field variations are

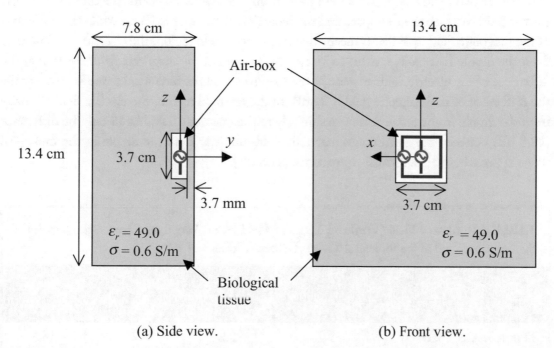

(a) Side view.   (b) Front view.

FIGURE 3.7: Wire antennas implanted in a simplified biological tissue model

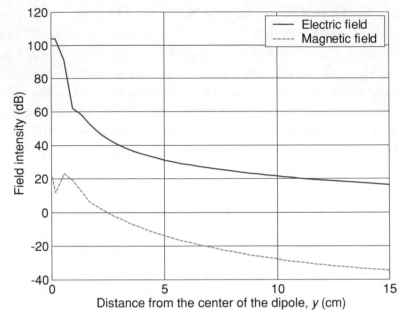

FIGURE 3.8: Field distributions along the y-axis from the small dipole in the biological tissue

observed at two locations, 5 and 15 cm away from the dipole. At 5 cm, the difference of the electric field between the free space and the biological tissue case is 28 dB while the difference of the magnetic field is 16 dB. It means that in the near-field region, the electric field intensity from the dipole in a biological tissue decreases faster than the magnetic field. At 15 cm, the difference of the electric field between in the free space and the biological tissue is 19 dB while the difference of the magnetic field is 20 dB. In the far-field region, the electric field intensity from the dipole in the tissue decreases similarly to the magnetic field. At 15 cm, the difference (48.8 dB) between the electric and magnetic field intensity from the dipole in the biological tissue approaches to the intrinsic impedance (51.5 dB).

TABLE 3.1: Electric Field (V/m) and Magnetic Field (A/m) Variations Between the Dipoles in the Free Space and in the Biological Tissue (Delivered Power = 1 W)

	5 cm		15 cm	
OBSERVATIONS	E (V/m dB)	H (A/m dB)	E (V/m dB)	H (A/m dB)
Dipole in free space	59.2	2.2	34.7	− 14.1
Dipole in biological tissue	31.2	−13.6	16.5	− 34.5

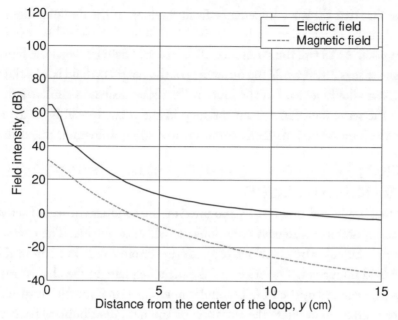

FIGURE 3.9: Field distributions along the y-axis from a square loop antenna in the biological tissue

Figure 3.9 shows the electric and magnetic field distributions along the y-axis from the loop antenna in the simplified tissue model. At the boundary between the tissue and the free space, abrupt variation in the slope of the electric field is observed while the magnetic field decreases continuously along the y-axis. The difference between the electric field and magnetic field increases as the distance increases from the antenna.

Table 3.2 shows the electric field and magnetic field variations along the y-axis from the loop. Similarly to dipole antenna, the field variations are observed at two locations, 5 and 15 cm from the center of the loop. At 5 cm, the difference of the electric field between the free space and the biological tissue case is 36 dB while that of the magnetic field is 18 dB. The electric field from the loop in the biological tissue decreases faster than the magnetic field as the distance

TABLE 3.2: Electric Field (V/m) and Magnetic Field (A/m) Variations Between the Loops in the Free Space and in the Biological Tissue (Delivered Power = 1 W)

	5 cm		15 cm	
OBSERVATIONS	E (V/m dB)	H (A/m dB)	E (V/m dB)	H (A/m dB)
Loop in free space	47.3	13.3	22.8	−10.7
Loop in biological tissue	11.3	−7.3	−3.1	−35.1

increases. The fact that the wave impedance from the loop in the tissue model is 18.6 dB at 5 cm indicates that the longitudinal magnetic field (H_y along the y-axis) is very strong in front of the tissue model. At 15 cm, the difference of the electric field between the free space and the biological tissue case is 26 dB while the difference of the magnetic field is 24 dB. In the far-field region, the electric field intensity of the loop in the tissue decreases similarly to the magnetic field. Because the wave impedance from the loop in the tissue model is 32 dB at 15 cm, it is expected that the longitudinal magnetic component is still dominant at this distance.

3.3   EFFECTS OF CONDUCTOR ON SMALL WIRE ANTENNAS IN BIOLOGICAL TISSUE

It is expected that implanted antennas are mounted on the conductive cases of active implantable medical devices in order to wirelessly communicate with the outside. The characteristic variations of the simple antennas by a conductive plate are estimated only in terms of the near electric and magnetic field intensities. The effects of the metallic plate on the characteristics of simple wire antennas are analyzed by the FDTD simulations. The same simulation structures as shown in Fig. 3.7 are utilized to evaluate the variation of the field distributions from the small wire antennas.

As shown in Fig. 3.10, a conductive plate which is parallel to the x–y plane is additionally included in the simulation structure behind the small wire antennas. The conductive plate can be considered as the surface of implantable medical devices. The wire antenna's input impedance

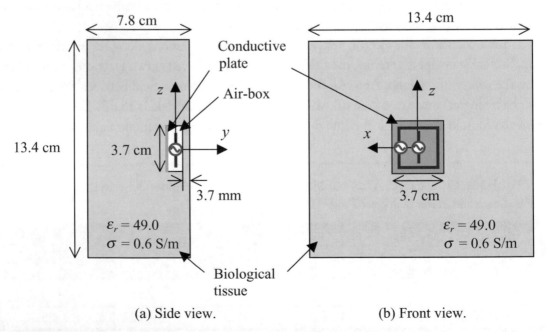

(a) Side view.   (b) Front view.

FIGURE 3.10: Wire antennas above conductive plate inside a simplified biological tissue model

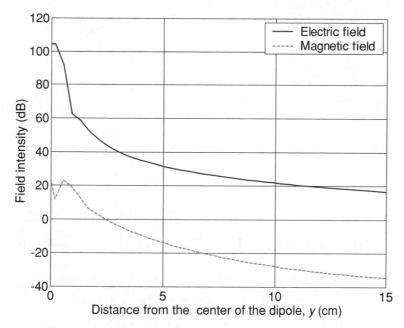

FIGURE 3.11: Field distributions along the y-axis from the small dipole in front of conductive plate in the biological tissue

is affected by the conductive plate. The impedance matching characteristic of the antenna is not considered here. The near-field variations from the dipole/loop antennas are the main focus in this chapter.

Figure 3.11 shows the electric and magnetic field distributions along the y-axis from the dipole antenna in front of the conductive plate in the simplified tissue model of Fig. 3.10. The dipole antenna delivers 1 W. Because the field distributions of Fig. 3.11 are very similar to those of Fig. 3.8, a detailed comparison between the two cases is required (Table 3.3).

From Table 3.3, it is observed that the conductive plate affects the electric field intensities by a slight increase of 0.3 dB along the y-axis although no variation in the magnetic field intensity is observed.

TABLE 3.3: Variations of Electric Field (V/m) and Magnetic Field (A/m) from the Dipole in the Biological Tissue by the Conductive Plate (Delivered Power = 1 W)

	5 cm		15 cm	
OBSERVATIONS	*E* (V/m dB)	*H* (A/m dB)	*E* (V/m dB)	*H* (A/m dB)
Without metal plate	31.2	− 13.6	16.5	− 34.5
With metal plate	31.5	− 13.6	16.8	− 34.5

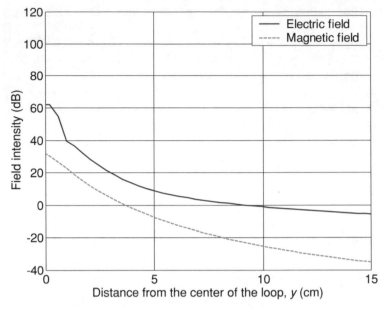

FIGURE 3.12: Field distributions along the y-axis from the small loop in front of conductive plate in the biological tissue

The electric and magnetic field distributions along the y-axis from the loop antenna in front of the conductive plate in the simplified tissue model are calculated in Fig. 3.12. The loop antenna delivers 1 W. The field distributions of Fig. 3.12 are very similar to those of Fig. 3.9 and a detailed comparison between the two cases is given in Table 3.4.

From Table 3.4, the electric field intensities vary from 11.3 to 8.9 dB (V/m) at 5 cm and from −3.1 to −5.5 dB (V/m) at 15 cm because of the metallic plate. The decreased values are the same as 2.4 dB at 5 and 15 cm. The fact that the magnetic field intensities are not changed signifies that the receiving power is not changed if the magnetic field from a loop antenna is coupled in the near-field region at the outside.

TABLE 3.4: Variations of Electric Field (V/m) and Magnetic Field (A/m) from the Loop in the Biological Tissue by the Conductive Plate (Delivered Power = 1 W)

OBSERVATIONS	5 cm		15 cm	
	E (V/m dB)	H (A/m dB)	E (V/m dB)	H (A/m dB)
Without metal plate	11.3	− 7.3	− 3.1	− 35.1
With metal plate	8.9	− 7.3	− 5.5	− 35.1

CHAPTER 4

Antennas Inside a Human Head

To ensure that applying simplified spherical head models for the characterization of implanted antennas in a head is adequate, electric field distributions from a dipole antenna in the spherical head model are compared with those in an anatomical head model. Three types of the spherical head models are used to assess how much the performance of implanted antennas depends on the head configurations. Based on the results of the dependency estimation, the maximum available power is calculated to give basic insights about the performance of the biomedical links built by implanted antennas in the spherical adult's and child's heads.

4.1   APPLICABILITY OF THE SPHERICAL HEAD MODELS

To reduce the discrepancies and increase the usefulness of the simplified spherical head models for the implanted antennas' characterization, the volume of the spherical head model should be matched to that of the anatomical head model. For the volume matching, the anatomical head model was scaled down from the original human phantom file of Fig. 2.5 in order to make the anatomical head's volume equal to the homogeneous spherical head's (radius = 9 cm, volume = 3.05×10^{-3} m^3) as shown in Fig. 4.1. The spherical head model of Fig. 4.1(a) for the spherical DGF simulations is composed of a single brain tissue. The anatomical head of Fig. 4.1(a) for the FDTD simulations consists of various biological tissues whose electrical characteristics are given in Table 4.1.

Figure 4.2 shows near electric field distributions calculated from the spherical DGF and FDTD codes. The dipole antennas (length = 5.3 cm) are positioned at the centers of the anatomical and spherical heads and deliver 1 W at 402 MHz. The electric field distribution differences inside the head between two codes are bigger than those outside the head due to complex biological tissues of the anatomical head model. However, because the implanted antenna inside a human head is analyzed for a wireless communication link, it is required to check the electric field difference outside a human head. Therefore, the fact that the largest near-field difference is 1.1 dB at 25 cm from the head's center provides enough evidence that the simplified spherical head model can be applied to characterize implanted antennas for biotelemetry applications instead of using more exact but complicated anatomical head models.

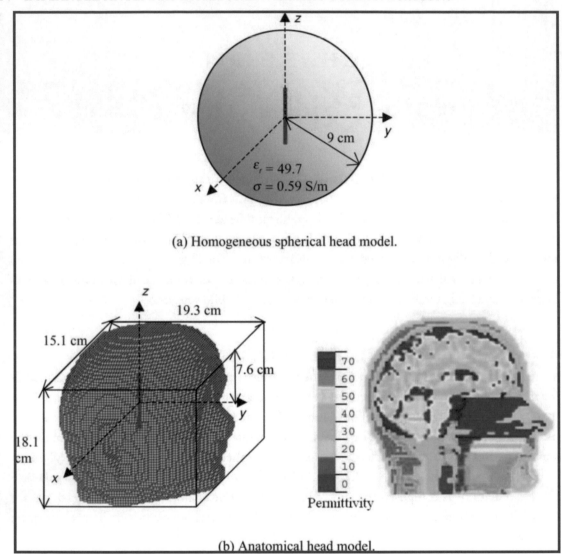

(a) Homogeneous spherical head model.

(b) Anatomical head model.

FIGURE 4.1: Volume-matched spherical head and anatomical head models

The condition for utilizing the spherical head model is that the volume of the spherical model should be matched to that of the anatomical head model.

4.2   ANTENNAS IN VARIOUS SPHERICAL HEAD MODELS

To check how much the electrical characteristics of implanted antenna rely on the spherical head's structure, half-wavelength ($0.5\lambda_d$) dipole antennas are positioned at two locations, 0

TABLE 4.1: Electrical Constants of Biological Tissues Used for the Anatomical Head Model at 402 MHz

BIOLOGICAL TISSUE	PERMITTIVITY (ε_r)	CONDUCTIVITY $(\sigma, \text{S/m})$
Brain	49.7	0.59
Cerebrospinal fluid	71.0	2.25
Dura	46.7	0.83
Bone	13.1	0.09
Fat	11.6	0.08
Skin	46.7	0.69
Skull	17.8	0.16
Muscle	58.8	0.84
Blood	64.2	1.35
Cartilage	45.4	0.59
Jaw bone	22.4	0.23
Cerebellum	55.9	1.03
Tongue	57.7	0.77
Mouth cavity	1.0	0.00
Eye tissue	57.7	1.00
Lens	48.1	0.67
Teeth	22.4	0.23

and 4.5 cm from the centers of the three kinds of heads (homogeneous, three-layered, and six-layered) as given in Table 2.2.

The near electric fields inside and outside the heads are calculated at 402 MHz and are compared in Fig. 4.3. The delivered power of a dipole is 1 W. Figures 4.3(a) and 4.3(b) show that the spherical head models have overall similar near electric field distributions. When the electric field values are closely compared, near-field differences are found to be largest at the edge of the heads due to different biological tissue components. Furthermore, smaller than 10% of the electric field intensity difference among the three heads is only observed at 30 cm away from the head center as long as the dipoles are located within 4.5 cm of the head centers.

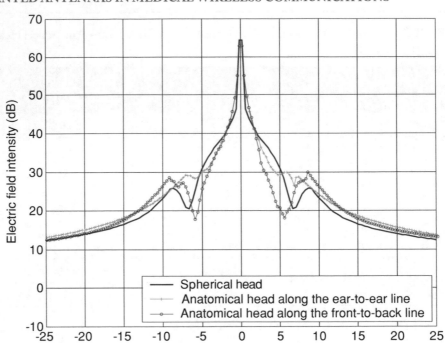

FIGURE 4.2: Near-field distributions for dipole antennas implanted at the centers of the volume-matched spherical and anatomical head models

Radiated powers from the half-wavelength dipole in the spherical head models (radius = 9 cm) at 402 MHz are compared in Table 4.2 when the delivered power is 1 W. Although the radiated powers for the three-layered head case are higher than those for other cases when the dipoles are located at the centers of the head models and 4.5 cm away from the centers, the differences are very small.

TABLE 4.2: Radiated Power from the Dipole in the Various Spherical Head Models (Delivered Power = 1 W)

HEAD MODELS	DIPOLE (0, 0, 0 cm) at 402 MHz	DIPOLE (0, 0, 4.5 cm) at 402 MHz
Homogeneous	14.7 mW	7.2 mW
Three-layered	16.6 mW	8.0 mW
Six-layered	13.4 mW	6.2 mW

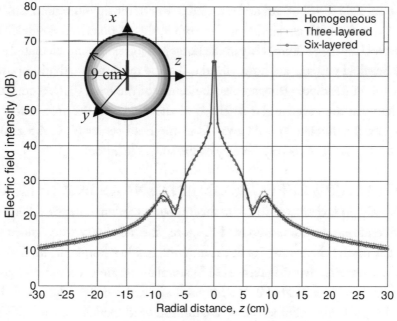

(a) Dipole's locations = (0, 0, 0 cm).

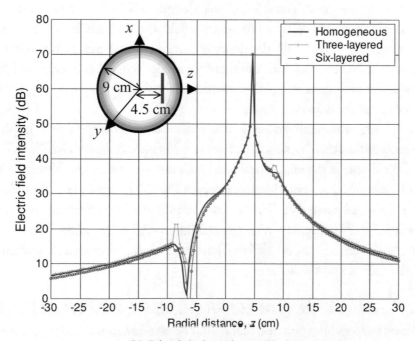

(b) Dipole's location = (0, 0, 4.5 cm).

FIGURE 4.3: Near-field distributions along the z-axis for dipole antennas implanted in the homogeneous, three-layered, and six-layered spherical head models

As the near-field distributions and radiated power of the dipole antenna inside three kinds of the spherical head models are similar to each other, the co-polarized far-field patterns of the implanted dipole (length = 5.3 cm) only in the homogeneous head are calculated at 402 MHz in Fig. 4.4. The far-field patterns are normalized using the maximum radiation power. As shown in Fig. 4.4(a), when the dipole is located at the center of the head (0, 0, 0 cm), the horizontal and vertical patterns are very similar to those of the dipole in the free space. As shown in Fig. 4.4(b), when the dipole is moved away from the head center (0, 0, 4.5 cm), the patterns become distorted because of the asymmetry of the source location.

4.3   SHOULDER'S EFFECTS ON ANTENNAS IN A HUMAN HEAD

In the previous chapter, characteristic data for implanted antenna inside a head are generated without consideration of a human body. Therefore, it is necessary to consider what effects a human body has on the characteristics of the antenna inside a human head.

To find the effects, two different FDTD simulations are executed and compared. One simulation is for the dipole located in the human head without shoulders, and the other is for the dipole in the head with shoulders. Two anatomical head models for the FDTD simulations are shown in Fig. 4.5. The only difference between the two simulation geometries is a 12 cm extended body below the neck. The $0.5\lambda_d$ dipole (length = 5.3 cm) antennas are located at the centers of two head models to calculate the electric field distributions from the antenna sources.

By comparing the near-field distribution from the head model without the shoulder with those from the head model with the shoulder in Fig. 4.6, it has been found that the field intensity outside the head decreases in the presence of the shoulder because the shoulder absorbs additional amount of the delivered power.

Figure 4.7 shows normalized horizontal radiation patterns for dipole implanted at the anatomical heads without/with a shoulder, based on Fig. 4.5. As shown in Fig. 4.7, the pattern differences of the dipole in the anatomical head without/with a shoulder become larger and the horizontal polarization (cross-pol.) level increases in Fig. 4.7 because of the shoulder. Additionally, when the delivered power is 1 W, the radiated power of the antenna in the head without the shoulder is 5.7 mW while that with the shoulder is 2.5 mW. Therefore, if an antenna is implanted in a real human head, the radiated patterns are more distorted than calculated in this chapter and the radiated power is lower.

4.4   ANTENNAS FOR WIRELESS COMMUNICATION LINKS

Figure 4.8 shows a communication link that is built between an implanted dipole and an exterior dipole. Two dipole antennas transmit the power and receive the power alternately in order to exchange the required data and information. For characterization of the link, dipole antennas are implanted in two different homogeneous heads to consider adult's head and child's head.

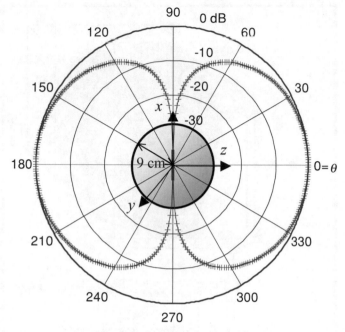

(a) Dipole's location = (0, 0, 0 cm).

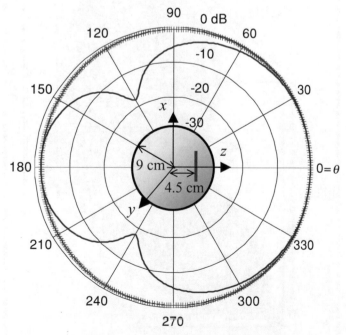

(b) Dipole's location = (0, 0, 4.5 cm).

FIGURE 4.4: Normalized radiation patterns for dipole antennas implanted in the homogeneous head model. (*Notes.* — Co-pol. (vertical pol.) over y–z plane, Co-pol. over x–z plane)

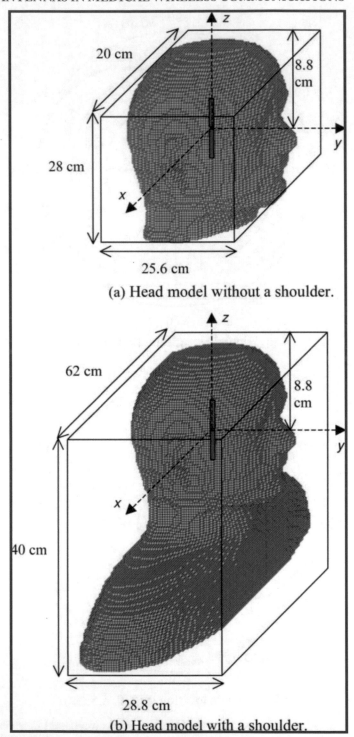

(a) Head model without a shoulder.

(b) Head model with a shoulder.

FIGURE 4.5: Dipole antennas located at the centers of anatomical head models without/with a shoulder

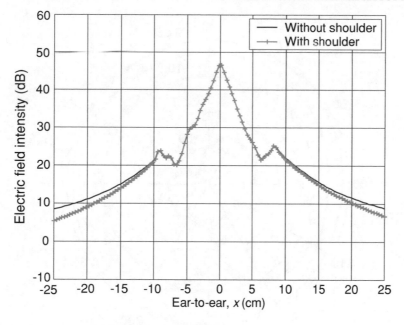

(a) Electric field distributions along x-axis.

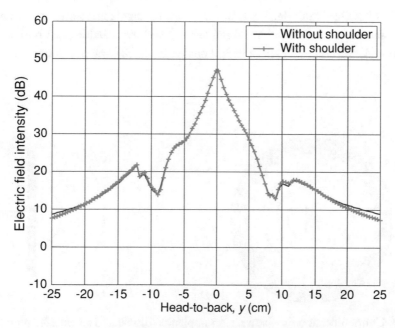

(b) Electric field distributions along y-axis.

FIGURE 4.6: Near-field distributions for dipole antennas in anatomical heads without/with a shoulder

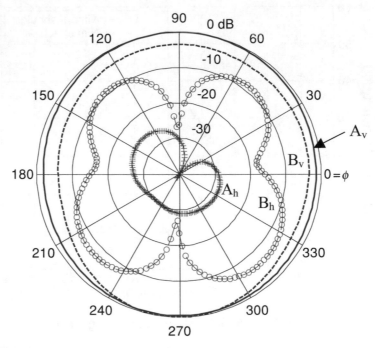

FIGURE 4.7: Normalized horizontal radiation patterns for dipole implanted in the anatomical heads without/with a shoulder. (*Notes.* A_v: vertical pol. pattern without shoulder, A_h: horizontal pol. pattern without shoulder, B_v: vertical pol. pattern with shoulder, B_h: horizontal pol. pattern with shoulder)

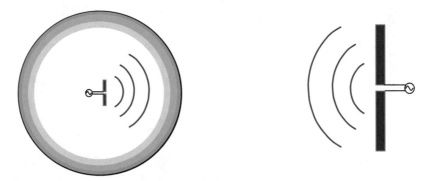

FIGURE 4.8: Communication link between an implanted dipole and an exterior dipole

It is assumed that an adult head is 9 cm in radius and a child head is 5 cm in radius. The homogenous heads consist of a single brain tissue.

To estimate the performance of the communication links, the maximum available power, P_{max} (W), is obtained from [21]:

$$p_{max} = S_{rec}A_{em} = S_{rec}\frac{\lambda^2}{4\pi}D_m \qquad (4.1)$$

where S_{rec} is the power density (W/m^2) at the receiving antenna, A_{em} the maximum effective aperture of the receiving antenna (m^2), λ the wavelength of the incoming wave, and D_m the maximum directivity of the receiving antenna (half-wavelength dipole = 1.64, infinitesimal dipole = 1.5).

In Fig. 4.9, a communication link is established between two dipole antennas. One dipole is 5.3 cm in length ($0.5\lambda_d$ – half dielectric wavelength) and implanted at the center of a spherical homogeneous head. The other is a half-wavelength ($0.5\lambda_0$ – half free-space wavelength) dipole and located in the free space. The implanted dipole transmits 1 W and the exterior dipole whose maximum directivity is 1.64 receives power within 5 m from the head center. The maximum available power at the exterior dipole is calculated when the implanted dipole is in different head sizes: 9 cm radius head and 5 cm radius head. A 9 cm radius head represents an adult head and a 5 cm radius head does a child head. When the exterior antenna is located within 5 m away from the adult head (9 cm radius), the maximum available power is higher than

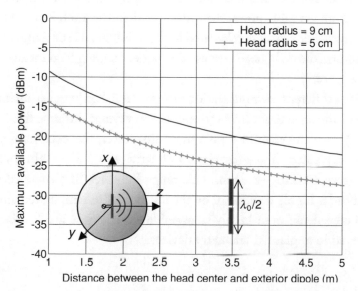

FIGURE 4.9: Maximum available power calculated at the exterior dipole when the implanted dipole delivers 1 W

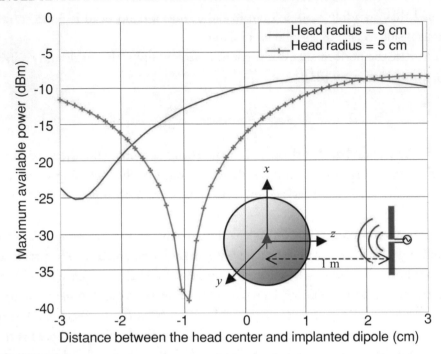

FIGURE 4.10: Maximum available powers calculated at the implanted dipole when the outside dipole transmits 1 W

−23 dBm. When the exterior antenna is located within 5 m away from the child head (5 cm radius), the maximum available power is 8 dB lower than the adult head. The reason that the higher maximum available power is calculated for the adult head is explained by higher fringing field generated around the child head because the higher fringing field means the lower radiated power in the free space.

In Fig. 4.10, a different communication link is established between an infinitesimal dipole in a spherical head and an exterior $0.5\lambda_0$ (free-space wavelength) dipole fixed at 1 m away from the head center. When the exterior dipole transmits power, the maximum available power is calculated at the implanted dipole (maximum directivity = 1.5) within 3 cm from the head center which is from −38 to −8 dBm. This range indicates that the electric field intensity inside a human head is highly dependent on the location of the implanted antenna. Because of this fact, careful consideration of highly changeable electric field inside a head is required for the receivers located in implanted telemetry devices.

CHAPTER 5

Antennas Inside a Human Body

In this chapter, antennas implanted in a human body are analyzed by the FDTD simulations for two main purposes. One purpose is to characterize a wire antenna which is located in a human internal organ, and the other is to design planar antennas for active medical devices which are located under a skin tissue.

5.1 WIRE ANTENNA INSIDE A HUMAN HEART

It is expected that antennas are implanted in a human body as a part of artificial organs such as a heart, kidney, etc. for potential medical applications. As a representative example, a short dipole is implanted inside the heart in a human body and the electromagnetic characteristics are calculated using the FDTD simulations.

Figure 5.1 shows that a human torso is generated from a whole human body in order to reduce the complexity and the time necessary for the FDTD simulations. Because the short dipole whose length is 1.1 cm is located inside a human heart, the dipole is 7.6 cm away from the free space. The dimension of the torso and location of the dipole are specified in Fig. 5.1. To overcome the direct contact between the antenna and biological tissues, the short dipole is shielded by a lossless dielectric cylinder whose length is 1.2 cm and radius is 0.4 cm.

Figure 5.2 shows the near and far electric field distributions generated from the short dipole whose delivered power is 1 W. As shown in Fig. 5.2(a), because the dipole is located near the chest, higher electric field intensities are observed in the positive y-axis (front body) than in the negative y-axis. Similar trend is observed in the normalized far-field pattern of Fig. 5.2(b) because about a 9 dB front-to-back lobe-ratio in the co-pol. far-field pattern is calculated. This means that positioning the exterior instrument in front of the chest is advantageous for the biomedical link when a wireless link between the dipole inside the human heart and the exterior instrument is considered. Additionally, co-pol. and cross-pol. ratio is about 20 dB in the boresight direction (y-axis).

5.2 PLANAR ANTENNA DESIGN

According to [12], implantable medical devices such as pacemakers and defibrillators are normally positioned under a skin tissue in a left upper human chest. Two types of low-profile

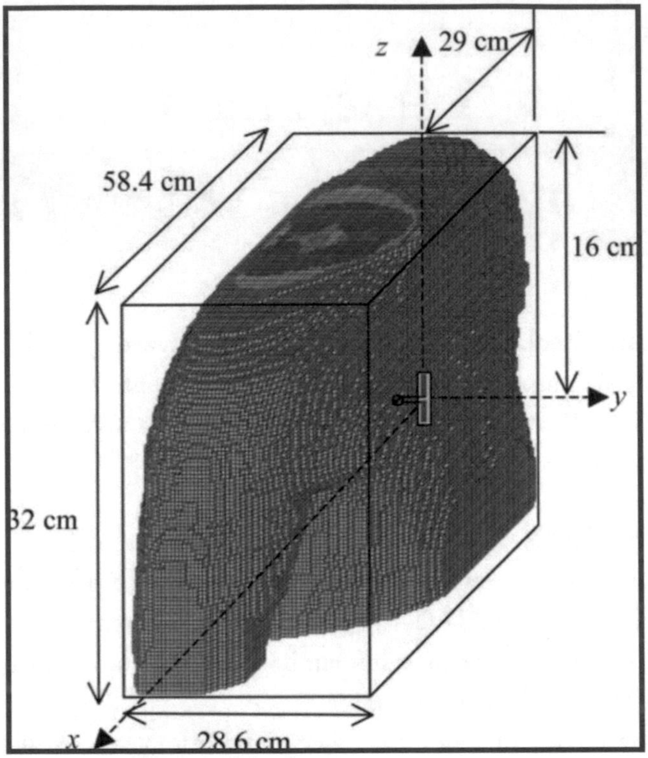

FIGURE 5.1: FDTD simulation geometry for the dipole antenna inside a human heart

antennas are considered to establish a communication link by mounting the antennas on implantable medical devices. One type is a microstrip antenna and the other is a planar inverted F antenna. For those antennas, spiral-type radiators are applied to reduce the total antenna size and additional dielectric layer (superstrate) is used on the spiral metallic radiator. Since superstrate dielectric layer is used, the metallic radiator does not directly come in contact with the surrounding biological tissues. Therefore, a superstate layer facilitates implanted antenna design by providing stable impedance matching performance of implanted antennas and lower absorbed power inside a human body.

Planar antennas are designed using the human torso FDTD geometry of Fig. 5.3. The origin of the coordinate system is located at the center of the geometry. Planar antennas are implanted under a skin tissue and are located 20 cm from the top of the geometry and 26 cm from the left end.

5.2.1 Microstrip Antenna

Figure 5.4 shows the shape and return-loss of a microstrip antenna designed in the FDTD human torso. The uniformly wide radiator of the microstrip antenna is a spiral type to reduce

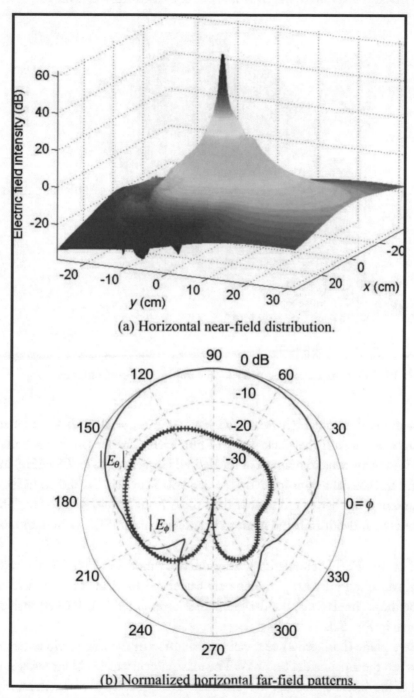

(a) Horizontal near-field distribution.

(b) Normalized horizontal far-field patterns.

FIGURE 5.2: Electric field distributions for the small dipole located inside the human heart

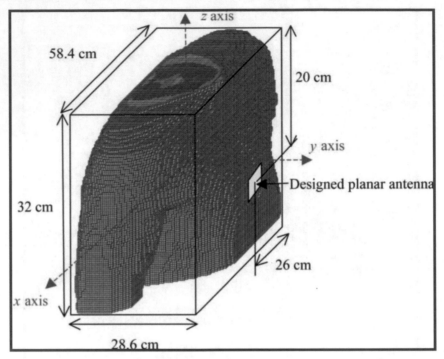

FIGURE 5.3: FDTD human torso geometry for the implanted planar antenna design

the antenna dimension and high permittivity material ($\varepsilon_r = 10.2$) is used as the substrate and superstrate layers with the thickness of 4 mm each. The length of the spiral radiator is determined for the antenna to resonate at the desired frequency (402–405 MHz) in the FDTD geometry and the coaxial feed is located for a good 50 Ω match. As shown in Fig. 5.4(b), the microstrip antenna has better return-loss than 6 dB from 397 to 423 MHz. Therefore, the 6 dB return-loss bandwidth for the microstrip antenna in the FDTD human model is about 6.3%.

Based on the FDTD simulations using the body model of Fig. 5.3, the radiation characteristics of the spiral microstrip antenna are calculated in terms of near-field and far-field patterns. For these simulations, the designed microstrip antenna is located inside the human chest as shown in Fig. 5.3.

The x–y plane (horizontal) near-field distribution of the microstrip antenna is given in Fig. 5.5(a) when the antenna delivers 1 W. The maximum electric field intensity is calculated at the location of the antenna. Because the planar antenna is located near the chest, higher electric field intensities are observed along the positive y-axis (front body) than along the negative y-axis. The normalized x–y plane (horizontal) far-field patterns of the antenna are given in Fig. 5.5(b). The maximum directivity is observed in the front of the human body. The E_θ power

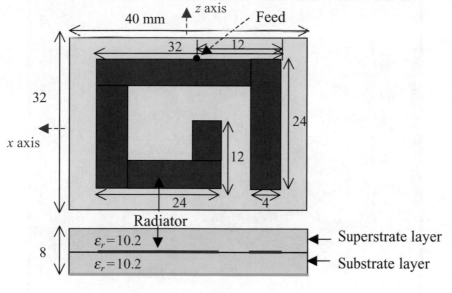

(a) Designed spiral microstrip

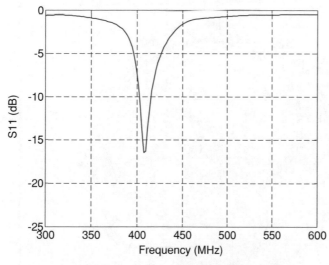

(b) Return-loss of spiral microstrip

FIGURE 5.4: Spiral microstrip antenna in the anatomical body model

levels of the antennas are similar to the E_ϕ levels because of the spiral radiator and the effects of the complex human body. Because the planar antenna is designed on a small ground plane for compact dimensions, the front-to-back lobe ratios of the antenna in the human chest are lower than 5 dB.

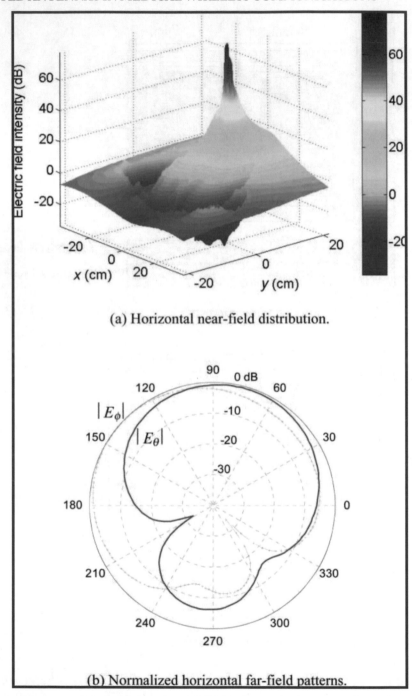

(a) Horizontal near-field distribution.

(b) Normalized horizontal far-field patterns.

FIGURE 5.5: Radiation characteristics of the spiral microstrip antenna in the anatomical body model

5.2.2   Planar Inverted F Antenna

Figure 5.6 shows the shape and return-loss of a planar inverted F antenna (PIFA) designed in the FDTD human torso. The PIFA has substrate and superstrate layers whose dielectric constants ($\varepsilon_r = 10.2$) and total thickness (8 mm) are the same as the spiral microstrip antenna. To achieve smaller dimension (24 mm × 32 mm) than the microstrip antenna, the shape of the

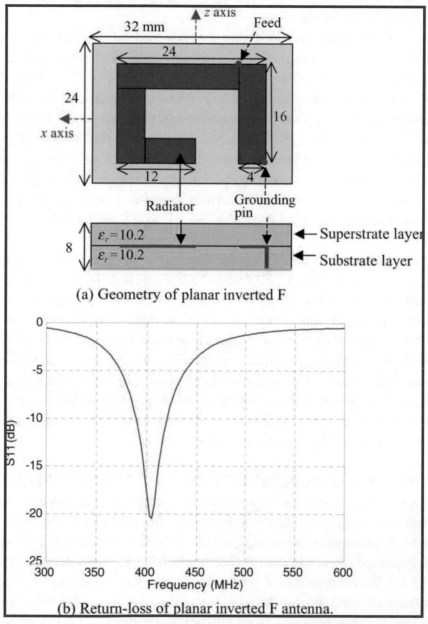

(a) Geometry of planar inverted F

(b) Return-loss of planar inverted F antenna.

FIGURE 5.6: Spiral planar inverted F antenna inside the anatomic body model radiator is spiral and a grounding pin is additionally located at the end of the radiator in order to connect the radiator to the ground plane. The length of the spiral radiator is determined such that the antenna to resonate at the desired frequency (402–405 MHz) in the FDTD geometry and the coaxial feed is located for a good 50 Ω match. The PIFA has better return-loss than 6 dB from 378 to 433 MHz. Therefore, the 6 dB return-loss bandwidth for the PIFA in the FDTD human body model is 13.6%. The bandwidth of the PIFA is more than double the microstrip antenna's bandwidth.

Based on the FDTD simulations using the body model of Fig. 5.3, the radiation characteristics of the PIFA are calculated in terms of near-field and far-field patterns. The x–y plane (horizontal) near-field pattern of the antenna is given in Fig. 5.7(a) when the antenna delivers 1 W. The near-field distributions of the PIFA are very similar to those of the microstrip antenna. The normalized x–y plane (horizontal) far-field patterns of the PIFA are calculated in Fig. 5.7(b). The maximum directivity is observed in the front of the human body, as expected. Although the patterns of the PIFA are similar to those of the microstrip antenna, the E_θ power level of the PIFA is a little higher than the E_ϕ level in the boresight direction ($\phi = 90$). Therefore, it is expected that a linear polarized antenna should be located straightly in front of a human chest in order to receive the θ-polarized electric field from the PIFA implanted inside a human chest.

The radiation power of the PIFA inside the chest of the human body is 2.5 mW while that of the microstrip antenna is 1.6 mW when both antennas deliver 1 W. In addition to the physical small size of the PIFA, the radiation efficiency of the PIFA is higher than that of the microstrip antenna. When radiation mechanisms of two different type antennas are compared, it is found that the microstrip antenna generates high electric fields, while the PIFA generates high electric fields as well as high electric currents which flow from the feed to the grounding pin. The absorbed power equation in the conducting body ($P_{abs} = \frac{1}{2} \int \sigma |E|^2 \, dV$), where σ is conductivity and $|E|$ electric field intensity, in the conducting body indicates that the absorbed power is related to the electric field, it is expected in the same lossy medium that a PIFA has higher radiation efficiency than a microstrip antenna.

5.3   WIRELESS LINK PERFORMANCES OF DESIGNED ANTENNA

As shown in Fig. 5.8, two communication links are established to compare the performance of an implanted communication link with that of the free-space link. The implanted link is between two $0.5\lambda_0$ (free-space wavelength) dipole in the free space and the free-space link is between an implanted antenna in a human body and a $0.5\lambda_0$ dipole in the free space.

In Fig. 5.9, the performances of two communication links are estimated when the transmitting antennas deliver 1 W and the exterior receiving dipoles are located between 20 and 30 cm away from the transmitting antennas. For the implanted antenna, the designed microstrip

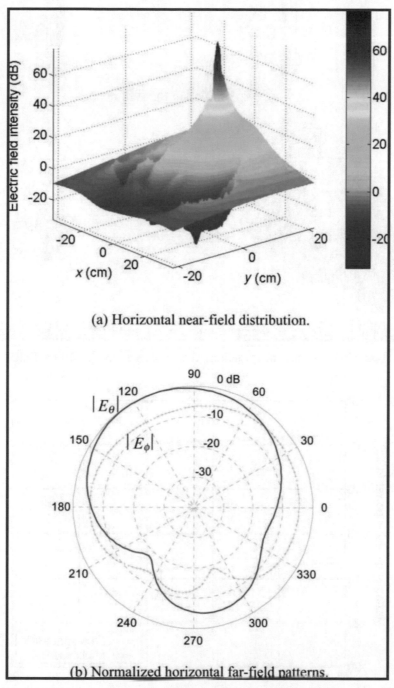

(a) Horizontal near-field distribution.

(b) Normalized horizontal far-field patterns.

FIGURE 5.7: Radiation characteristics of the PIFA in the anatomical body model

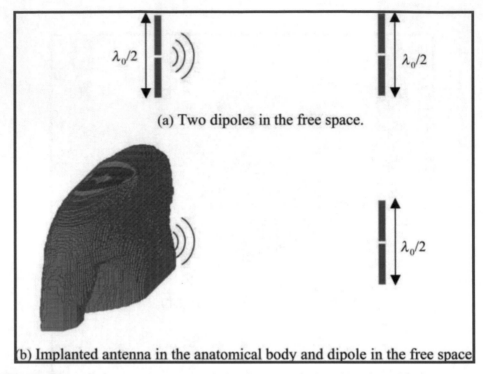

(a) Two dipoles in the free space.

(b) Implanted antenna in the anatomical body and dipole in the free space

FIGURE 5.8: Two wireless communication links: free-space link and implanted link

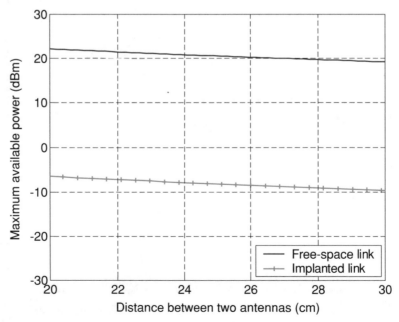

FIGURE 5.9: Maximum available power at the exterior dipole both from the dipole in free space and from the microstrip antenna in the anatomical body model when both antennas deliver 1 W

antenna are located in the FDTD human torso model of Fig. 5.3. Using Eq. (4.1), the maximum available power (dBm) are calculated using the power density (W/m^2) at the dipole in a free space, the wavelength (0.704 m at 402 MHz) of the incoming wave, and the maximum directivity (1.64) of the dipole antenna. The maximum available power difference (28 dB) between two links can be explained by the radiation efficiency ($10\times \log(0.16$ mW/1 W$) = -30$ dB) of the microstrip antenna in the FDTD human torso model of Fig. 5.3 and the fact that two transmitting antennas' far-field patterns are different from each other.

The maximum available power calculated outside the dipole is useful to estimate the required sensitivity of the outside receiver. For example, from the above calculation, the exterior receiver even within 0.3 m from the implanted antenna should have a front-end receiver whose sensitivity is better than about -10 dBm to established a reliable communication link when the implanted antenna transmits 1 W. Actually, the transmitted power, 1 W is used for the normalization purpose because the real transmitted power should be much lower than 1 W in order to satisfy the peak spatial-average SAR or other regulations related to active implantable medical devices. In the next chapter, the possible transmitted (delivered) power from implanted antennas will be calculated in consideration of the SAR regulation.

CHAPTER 6

Planar Antennas for Active Implantable Medical Devices

Compact planar antennas are designed, constructed and measured using finite difference time domain (FDTD) simulations and measurement setup for active implantable medical devices at the medical implant communications service (MICS) frequency band, 402–405 MHz. A planar inverted F antenna (PIFA) structure is applied to design two small low-profile antennas: meandered-type and spiral-type PIFA. The measurement setup is built by using a tissue-simulating fluid to make return-loss experiments on the constructed antennas. After the designed antennas are mounted on a medical device, the input impedance variation of both antennas is calculated by FDTD. The characteristics of both antennas are compared in terms of radiation performances and safety issue related to active implantable medical devices.

6.1 DESIGN OF PLANAR ANTENNAS
6.1.1 Simplified Body Model and Measurement Setup

For the ease of designing implanted antennas, planar antennas are located inside a simplified body model instead of an anatomic complete body model such as in Fig. 5.3. Because implantable medical devices are positioned under a skin tissue, electrical effects of the skin tissue on implanted antennas are very strong, a simplified body model consists of only one skin tissue (dielectric constant $(\varepsilon_r) = 46.7$, conductivity $(\sigma) = 0.69$ S/m at 402 MHz, mass density $(\rho) = 1.01$ g/cm^3), as shown in Fig. 6.1. The dimension of the hexahedral body model is 10 cm × 10 cm × 5 cm, and planar antennas are positioned at the center of the body model while the location of the antenna from the bottom of the body model is 1 cm.

The resonant characteristics of the designed antenna are measured in a human tissue-simulating fluid which was made from deionized water, sugar, salt, cellulose, etc. [22], as shown in Fig. 6.2. The electrical characteristics of the fluid ($\varepsilon_r = 49.6$, $\sigma = 0.51$ S/m at 402 MHz) are very similar to the skin tissue of the simplified body model. For the measurement of the return-loss characteristics, the antennas were positioned in the container filled with the fluid. The distance between the antenna and the bottom of the fluid is the same as the distance in the body model used for FDTD simulations.

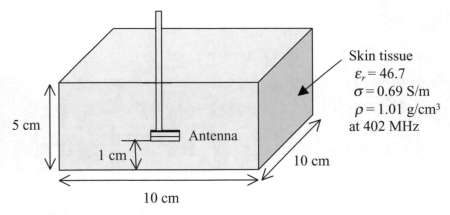

FIGURE 6.1: Simplified body model for the design of planar antennas implanted in a human body

6.1.2   Meandered PIFA

As shown in Fig. 6.3, a meandered antenna is designed for implantable medical device inside a human body at a biomedical frequency range of 402–405 MHz. Because the designed antenna uses a grounding pin at the end of the radiator, the operation mechanism is the same as a planar inverted F antenna (PIFA). The printed radiator is located between substrate and superstrate

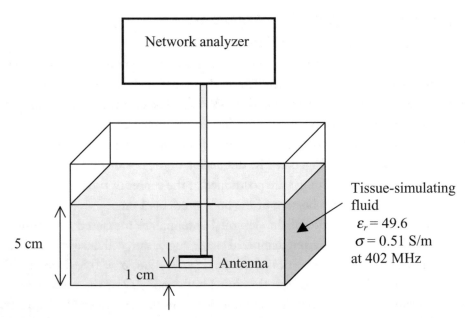

FIGURE 6.2: Return-loss measurement setup for constructed planar antennas using tissue-simulating fluid

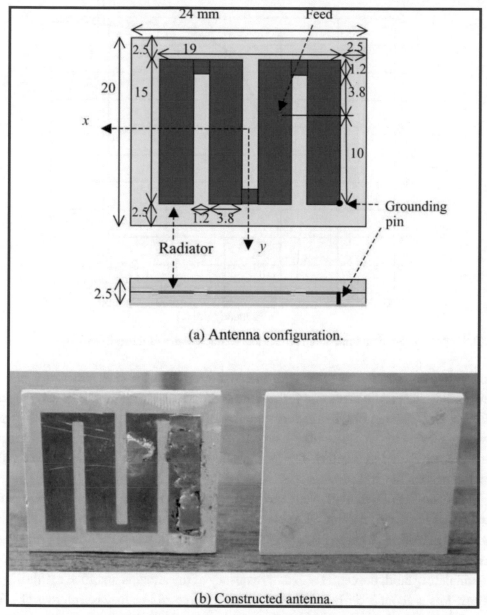

(a) Antenna configuration.

(b) Constructed antenna.

FIGURE 6.3: Meandered PIFA designed for the implantable device inside the simplified body model

dielectric layers whose dielectric constant is 10.2 and thickness is 1.25 mm. The origin of the coordinate system is located at the center of the ground plane which is 24 mm in width and 20 mm in length.

To understand the construction method of the meandered PIFA in Fig. 6.3, it is considered that the meandered radiator consists of four rectangular strips (15 mm × 3.8 mm) which are

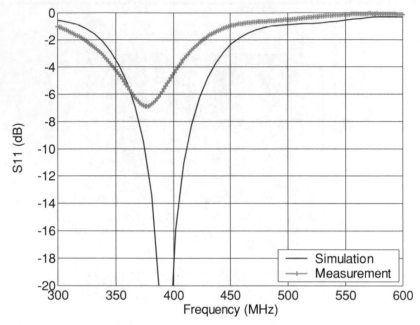

FIGURE 6.4: Simulated and measured return-loss characteristics of meandered PIFA

electrically connected to each other with three connection strips (1.2 mm × 1.2 mm). The spacing among the rectangular strips is the same as the distance (1.2 mm) between the radiator and the ground plane in order to reduce coupling effects between the rectangular strips and achieve a small antenna. By changing the length of the connection strips, the resonant frequency of the meandered antenna is tuned. The location of a coaxial feeding is determined to make the antenna match well to 50 Ω systems.

Using the FDTD simulation and measurement setups of Figs. 6.1 and 6.2, the matching characteristics for the meandered PIFA are compared in Fig. 6.4. The meandered PIFA shows a good 50 Ω matching characteristic at the desired frequency (402–405 MHz) in the simulated results. However, when the constructed antenna is positioned at 1 cm from the bottom of the tissue-simulating fluid, the center resonant frequency of the antenna is shifted a little down and the return-loss is about 3–5 dB at 402–405 MHz. The return loss may be improved by further tuning of the antenna.

6.1.3   Spiral PIFA

A spiral PIFA antenna is shown in Fig. 6.5. The uniform width radiator is sandwiched between substrate and superstrate dielectric layers whose thickness is 1.25 mm each and dielectric constant is 10.2. The origin of the coordinate system is located at the center of the ground plane (24 mm × 20 mm). Similar to the meandered PIFA, the spacing among the metallic strips

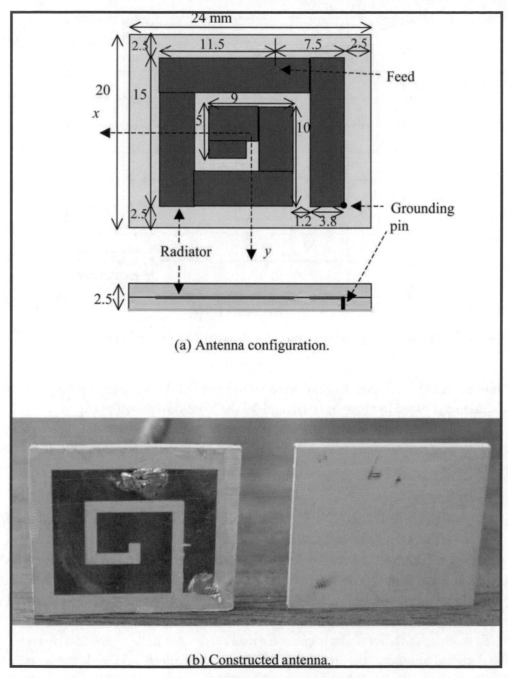

(a) Antenna configuration.

(b) Constructed antenna.

FIGURE 6.5: Spiral PIFA designed for the implantable device inside a human body

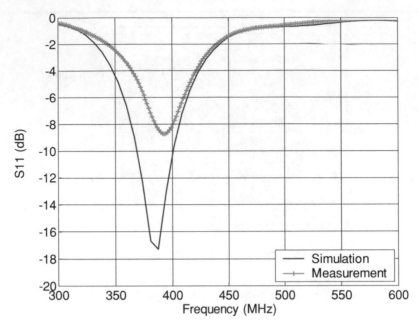

FIGURE 6.6: Simulated and measured return-loss characteristics of spiral PIFA

(3.8 mm in width) is 1.2 mm. In contrast to meandered PIFA, the operating frequency of the spiral antenna is tuned by changing the length of the innermost metallic strip.

In Fig. 6.6, when the spiral PIFA is positioned at 1 cm from the free space, the simulated matching performance (about 7–10 dB return-loss) is similar to the measured one at 402–405 MHz.

6.2   ANTENNA MOUNTED ON IMPLANTABLE MEDICAL DEVICE

6.2.1   Effects of Implantable Medical Device

For providing wireless communication links, an antenna is mounted on an implantable medical device, as shown in Fig. 6.7 based on Fig. 6.1. The implantable device is simulated by a metallic box made of six-sided conducting plates. The coaxial feeding system which is composed of a source and an absorbing boundary [23] is located inside the metallic box.

To estimate the effects of an implantable medical device on the return-loss characteristics of the designed antennas, the input impedance values of the PIFAs with the metallic box are compared with those of the antennas without the metallic box. The antennas mounted on the metallic box are inserted in the simulation model of Fig. 6.1. Because of the metallic box, the zero-crossing frequencies (resonant frequencies) of the imaginary impedances in the meandered and spiral PIFA cases are shifted down 1.5 and 1.3%, respectively. Additionally, the

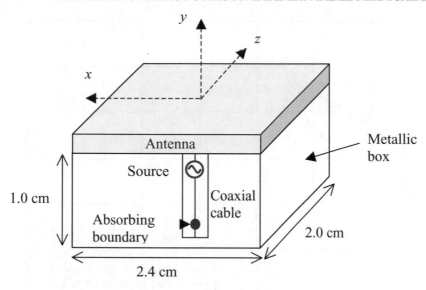

FIGURE 6.7: Antenna mounted on an implantable medical device

small change in the real and imaginary input-impedances indicates that the overall effects of implantable medical devices on the implanted PIFAs are negligible.

6.2.2 Near-Field and SAR Characteristics of Designed Antennas

After mounting the designed meandered and spiral PIFA on the metallic box in Fig. 6.7, near electric field and 1-g averaged SAR distributions are calculated for the antennas which are located at 1 cm from the free space in the simplified body model (Fig. 6.1). The near electric field distributions are calculated in x–z plane in front of the antennas ($y = 1.25$ mm). By following the numerical computational procedures recommended by IEEE [14], the SAR distributions for two antennas are given at $y = 0.5$ cm over x–z plane. The SAR value at each point is averaged using a 1 cm × 1 cm × 1 cm cube whose mass is almost 1 g because the mass density of the biological tissue is 1.01 g/cm^3.

Figure 6.9 shows the near electric field and 1-g SAR distributions of the meandered planar antenna when the antenna delivers 1 W. In the near-field distribution, the peak electric field intensity is observed at the end strip of the meandered radiator because the electric field intensity is maximum at the open end of a planar inverter F antenna. According to the 1-g SAR distribution of Fig. 6.9(b), the peak SAR value (24.7 dB = 294 mW/g) for the meandered PIFA is recorded in front of the left side of the radiator ($x = 6.3$, $z = 3.8$ mm) due to the peak electric field intensity.

Figure 6.10 shows the near electric field and 1-g SAR distributions of the spiral planar antenna when the antenna delivers 1 W. In the near-field distribution, the peak electric field

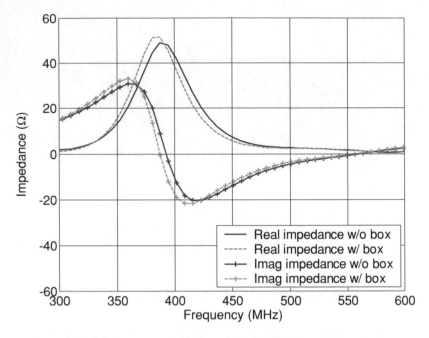

(a) Meandered PIFA.

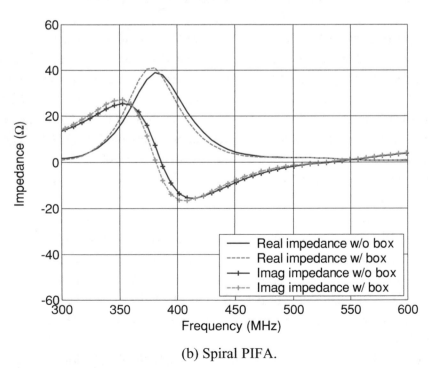

(b) Spiral PIFA.

FIGURE 6.8: Input impedance variations of the implanted antennas without/with the metallic box

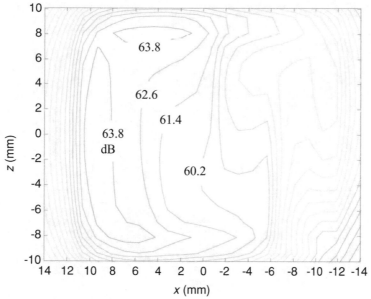

(a) Near-field distribution (0 dB = 20 × log (1 volt/m)).

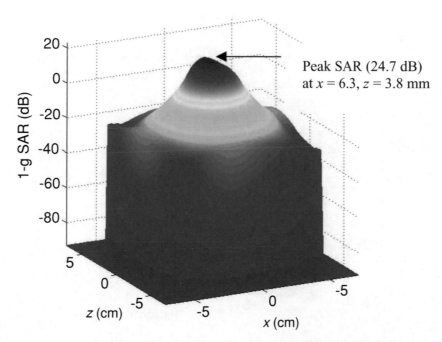

(b) 1-g SAR distribution (0 dB = 10 × log (1 m/Wg)).

FIGURE 6.9: Near electric field distribution and 1-g SAR distribution for the meandered PIFA (delivered power = 1 W)

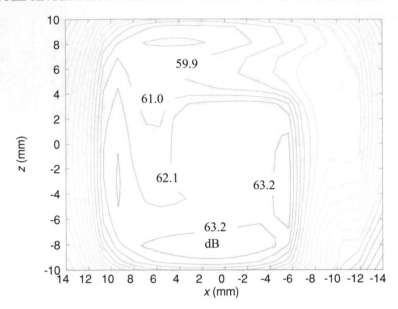

(a) Near-field distribution (0 dB $= 20 \times \log (1$ volt/m$))$.

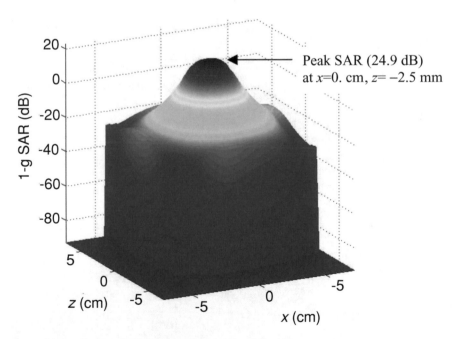

(b) 1-g SAR distribution (0 dB $= 10 \times \log (1$ m W/g$))$.

FIGURE 6.10: Near electric field distribution and 1-g SAR distribution for the spiral antenna (delivered power $= 1$ W)

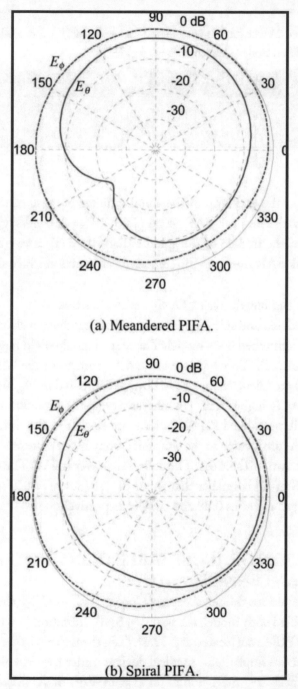

(a) Meandered PIFA.

(b) Spiral PIFA.

FIGURE 6.11: Comparison of $x–y$ plane radiation patterns between the meandered and spiral PIFA located at 2.5 mm from the bottom of the simplified body model

TABLE 6.1: Radiated Power for the Meandered and Spiral PIFA Located at 2.5 mm from the Bottom of the Simplified Body (Delivered Power = 1 W)

	RADIATED POWER (mW)
Meandered PIFA	2.8
Spiral PIFA	3.4

intensity is observed at the end strip of the spiral radiator because of the same reason as the meandered PIFA's case. The peak SAR value (24.9 dB = 310 mW/g) for the spiral PIFA is recorded in front of the middle of the spiral radiator ($x = 0$, $z = -2.5$ mm), as shown in Fig. 6.10(b). The peak SAR from two different-type antennas are very similar to each other.

6.2.3 Radiation Characteristics of Designed Antennas

To compare the meandered and spiral antennas in terms of radiation characteristics, two antennas are located in the simplified body model. The $x-y$ plane far-field radiation patterns for the meandered and the spiral PIFAs which are located 2.5 mm from the bottom of the simplified body model (Fig. 6.1) are calculated at 402 MHz and compared in Fig. 6.11. It is observed that $|E_\phi|$ pattern directivity is higher than $|E_\theta|$ directivity in the direction of $\phi = 90^0$ (boresight direction). The overall meandered PIFA's patterns are similar to the spiral PIFA's.

When the PIFAs are located in the simplified body model, the amount of power radiated in the free space are shown in Table 6.1. The delivered power is 1 W. The radiated powers of the meandered PIFA (2.8 mW) is smaller that spiral PIFAs (3.4 mW). Therefore, two antennas' radiation efficiencies are as low as 0.28 and 0.34%, respectively.

6.3 ESTIMATION OF ACCEPTABLE DELIVERED POWER FROM PLANAR ANTENNAS

To calculate the possible maximum acceptable power delivered by the designed implanted antennas in the simplified body model, the peak 1-g SAR limitation (1.6 mW/g = 1.6 W/kg) of ANSI is applied. The PIFAs are located at 2.5 mm from the bottom of the body model (Fig. 6.1) in order to consider active implantable medical devices under the skin biological tissue.

To satisfy the peak 1-g SAR limitation (1.6 mW/g) of ANSI, both antennas should not deliver 1 W because the peak SAR for the meandered and spiral PIFAs are 420 and 407 mW/g, respectively. From a simple calculation, the possible maximum delivered power from both antennas are 2.4 and 2.5 mW. By applying the radiation efficiencies (0.28% for the

TABLE 6.2: Radiated Power for Both PIFAs Located at 2.5 mm from the Bottom of the Simplified Body Model (Delivered Power = 2.4 and 2.5 mW from the Meandered and Spiral PIFA, Respectively)

	RADIATED POWER (mW)
Meandered PIFA	6.7 µW
Spiral PIFA	8.5 µW

meandered PIFA, and 0.34% for the spiral PIFA), the radiated power from both antennas to the free space are calculated as shown in Table 6.2.

It should be noted that the calculated possible maximum radiated power (6.7 and 8.5 µW for the meandered and spiral PIFA) in free space are even lower than the maximum effective radiated power (ERP) limitation (25 µW) of ERC [6].

CHAPTER 7

Conclusion

As mentioned in the Introduction, one of the main objectives of this book has been to summarize the results of recent research activities of the authors on the subject of implanted antennas for medical wireless communication systems. It is believed that ever sophisticated medical devices will be implanted inside the human body for medical telemetry and telemedicine. To establish effective and efficient wireless links with these devices, it is pivotal to give special attention to the antenna designs that are required to be low profile, small, safe and cost effective. In this book, it has been demonstrated how advanced electromagnetic numerical techniques can be utilized to design these antennas inside as realistic human body environment as possible. Also, it has been shown how simplified models can assist the initial designs of these antennas in an efficient manner.

To characterize and design implanted antennas inside a human head/body for biomedical wireless communications, two numerical methodologies—spherical dyadic Green's function (DGF) and finite difference time domain (FDTD)—are utilized. For the spherical DGF code, the human head is simplified as a lossy sphere which consists of different electrical layers. By introducing the infinitesimal current decomposition of the dipole antenna and the axis rotation, the general DGF expressions for the implanted antenna inside the multi-layered spherical structure are modified to implement the spherical DGF code. The law of conservation of energy is utilized not only to normalize the power delivered by the antenna in the spherical DGF code, but also to verify the DGF simulations. Furthermore, the results of the FDTD code is compared with those of the DGF code in order to give evidence for how well the two results match each other.

The electric performances of simple wire antennas such as dipole and loop antennas in a simplified biological tissue model are compared in the near-field region. Because the tissue model is lossy, the near-field intensities from the antennas decrease more rapidly than the antennas in the free space. It is very useful to know which field component is dominant in the near-field region in order to couple a maximum energy from the transmitting antennas. The effects of the conductive plate on the wire antennas are analyzed to estimate the characteristics variations of implanted antennas mounted on the cases of active medical devices.

The characteristics of a dipole implanted inside a human head are analyzed by comparing the results of the spherical DGF expansions with those of the FDTD techniques. The near- and far-field distributions obtained from the DGF and FDTD codes are useful not only for understanding the properties of the implanted antenna and performing parametric studies, but also for estimating how accurately the DGF code is able to produce the characteristic data of the dipole antenna inside a human head.

The FDTD simulation results additionally show that a shoulder has a larger impact on the field outside the head than the field inside the head when the dipole is located at the center of the head. Differences in the horizontal radiation patterns between the structure without a shoulder and the structure with a shoulder were also observed. As a result, we recommend that a large portion of the human body (neck, shoulder, etc.) should be included in the FDTD simulation geometry to obtain correct field distributions outside the head when the antenna is operating at the medical implantable communication service (MICS) frequency band of 402–405 MHz.

The resonant characteristics of the low-profile implanted antennas positioned in the left chest are optimized using the anatomic human body. Based on the FDTD simulations, a spiral-type microstrip and planar inverted F antennas (PIFA) at 402–405 MHz are designed in consideration of matching to the surrounding biological tissues. Although the radiation patterns are similar to each other, the PIFA has advantages over a microstrip antenna, specifically smaller dimensions and higher radiation efficiency. Additionally, maximum available power is calculated to estimate the performance of communication links between the designed antennas and exterior devices and can be used to anticipate how sensitive receivers are necessary for the reliable communication links. The maximum delivered power for both the antennas should be determined so that the SAR values of the antennas satisfy ANSI SAR limitations.

Through the FDTD simulation and experimental setups, low-profile PIFAs are designed and constructed for active implantable medical devices to communicate with exterior telemetry equipment. Meandered-shaped and spiral-shaped radiators are applied to reduce the overall antenna dimensions. For parametric studies of an implanted antenna, the FDTD simulations structures include a metallic box in order to configure an implantable metallic medical device on which the planar antennas are mounted. After the antennas are mounted on the metallic box, the small variation of the antenna's input impedance indicates that the effects of the medical device on the antenna's characteristics are negligible. By comparing the meandered and spiral PIFA in the simplified body model, it is found that the radiation performances of the spiral-shaped PIFA are similar to those of the meandered-shaped PIFA in terms of near-field, far-field, and specific absorption rate patterns.

Future research, engineering developments and medical advances will pave the way for effective and useful applications of implemented antennas in a variety of medical wireless communications systems.

References

[1]   C. H. Durney and M. F. Iskander, "Antennas for medical applications," in *Antenna Handbook*, Y. T. Lo and S. W. Lee, Eds. New York: Van Nostrand, 1988, ch. 24.

[2]   A. Rosen, M. A. Stuchly, and A. V. Vorst, "Applications of RF/microwaves in medicine," *IEEE Trans. Microwave Theory Tech.*, vol. 50, no. 3, pp. 963–974, Mar. 2002. doi:10.1109/22.989979

[3]   B. M. Steinhaus, R. E. Smith, and P. Crosby, "The role of telecommunications in future implantable device systems," in *Proc. 16th IEEE EMBS Conf.*, Baltimore, MD, pp. 1013–1014, 1994.

[4]   P. E. Ross, "Managing care through the air," *IEEE Spectrum*, pp. 26–31, Dec. 2004. doi:10.1109/MSPEC.2004.1363637

[5]   "Medical Implant Communications Service (MICS) Federal Register," *Rules and Regulations*, vol. 64, no. 240, pp. 69926–69934, Dec. 1999.

[6]   "ERC Recommendation 70-03 relating to the use of Short Range Devices (SRD)," *Eur. Postal Telecommun. Admin. Conf. CEPT/ERC 70-03*, Annex 12, Tromsφ, Norway, 1997.

[7]   C. Gabriel and S. Gabriel, "Compilation of the dielectric properties of body tissues at RF and microwave frequencies," Armstrong Laboratory, London, UK, Available at: http://www.brooks.af.mil/AFRL/HED/hedr/reports/dielectric/home.html.

[8]   C. T. Tai, *Dyadic Green's Functions in Electromagnetic Theory*, Scranton, PA: Intext Education, 1971.

[9]   L. Li, P. Kooi, M. Leong, and T. Yeo, "Electromagnetic dyadic Green's function in spherically multilayered media," *IEEE Trans. Microwave Theory Tech.*, vol. 42, no. 12, pp. 2302–2310, Dec. 1994.doi:10.1109/22.339756

[10] Y. Rahmat-Samii, K. W. Kim, M. Jensen, K. Fujimoto, and O. Edvardson, "Antennas and humans in personal communications," in *Mobile Antenna Systems Handbook*, K. Fujimoto and J. R. James, Eds., 2nd ed. Norwood, MA: Artech House, 2000, ch. 7.

[11] K. S. Nikita, G. S. Stamatakos, N.K. Uzunoglu, and A. Karafotias, "Analysis of the interaction between a layered spherical human head model and a finite-length dipole," *IEEE Trans. Microwave Theory Tech.*, vol. 48, pp. 2003–2013, Nov. 2000. doi:10.1109/22.884189

[12] D. Wessels, "Implantable pacemakers and defibrillators: device overview & EMI considerations," *IEEE Electromagn. Compat. Int. Symp.*, vol. 2, pp. 911–915, 2002.

[13] *IEEE Standard for Safety Levels with Respect to Human Exposure to Radio Frequency Electromagnetic Fields, 3 kHz to 300 GHz*, IEEE Standard C95.1-1999, 1999.

[14] *IEEE Recommended Practice for Measurements and Computations of Radio Frequency Electromagnetic Fields with Respect to Human Exposure to such Fields, 100 kHz to 300 GHz*, IEEE Standard C95.3-2002, 2002.

[15] K. W. Kim and Y. Rahmat-Samii, "EM interactions between handheld antennas and human: anatomical head vs. multi-layered spherical head," *IEEE Conf. Antennas Propagat. Wireless Comm.*, 1988.

[16] N. C. Skaropoulos, M. P. Ioannidou, and D. P. Chrissoulidis, "Induced EM field in a layered eccentric spheres model of the head: plane-wave and localized source exposure," *IEEE Trans. Microwave Theory Tech.*, vol. 44, pp. 1963–1973, 1996. doi:10.1109/22.539956

[17] K. W. Kim and Y. Rahmat-Samii, "Personal communication antenna characterization in the presence of a human operator," UCLA Report No. Eng-97-175, 1997.

[18] I. G. Zubal, C. R. Harrell, E. O. Smith, Z. Rattner, G. Gindi, and P. B. Hoffer, "Computerized three-dimensional segmented human anatomy," *Med. Phys.*, vol. 21, no. 2, pp. 299–302, Feb. 1994.

[19] O. P. Gandhi, G. Lazzi, and C. M. Furse, "Electromagnetic absorption in the human head and neck for mobile telephones at 835 and 1900 MHz," *IEEE Trans. Microwave Theory Tech.*, vol. 44, no. 10, pp. 1884–1897, Oct. 1996.doi:10.1109/22.539947

[20] C. A. Balanis, *Antenna Theory: Analysis and Design*, 2nd ed. John Wiley & Sons, 1997.

[21] W. L. Stutzman and G. A. Thiele, *Antenna Theory and Design*, 2nd ed. John Wiley & Sons, 1998.

[22] Application Note: Recipes for Head Tissue Simulating Liquids, Schmid & Partner Eng. AG, Zurich, Switzerland, 2002.

[23] M. A. Jensen and Y. Rahmat-Samii, "Performance analysis of antennas for hand-held transceivers using FDTD," *IEEE Trans. Antennas Propagat.*, vol. 42, no. 8, pp. 1106–1113, Aug. 1994.doi:10.1109/8.310002

Author Biographies

Yahya Rahmat-Samii received the M.S. and Ph.D. degrees in electrical engineering from the University of Illinois, Urbana-Champaign.

He is a Distinguished Professor and past Chairman of the Electrical Engineering Department, University of California, Los Angeles (UCLA). He was a Senior Research Scientist with the National Aeronautics and Space Administration (NASA) Jet Propulsion Laboratory (JPL), California Institute of Technology prior to joining UCLA in 1989. In summer 1986, he was a Guest Professor with the Technical University of Denmark (TUD). He has also been a consultant to numerous aerospace companies.

He has been editor and guest editor of numerous technical journals and books. He has authored and coauthored over 660 technical journal and conference papers and has written 20 book chapters. He coauthored *Electromagnetic Optimization by Genetic Algorithms (New York: Wiley, 1999) and Impedance Boundary Conditions in Electromagnetics* (New York: Taylor & Francis, 1995). He also holds several patents. He has had pioneering research contributions in diverse areas of electromagnetics, antennas, measurement and diagnostics techniques, numerical and asymptotic methods, satellite and personal communications, human/antenna interactions, frequency selective surfaces, electromagnetic band-gap structures, applications of the genetic algorithms and particle swarm optimization, etc., (visit http://www.ee.ucla.edu/antlab). On several occasions, his research has made the cover of magazines and has been featured on several TV News casts. He is listed in Who's Who in America, Who's Who in Frontiers of Science and Technology and Who's Who in Engineering. Professor Rahmat-Samii is the designer of the IEEE Antennas and Propagation Society (IEEE AP-S) logo, which is displayed on all IEEE-AP-S publications.

Dr. Rahmat-Samii is a member of Commissions A, B, J and K of USNC/URSI, Antenna Measurement Techniques Association (AMTA), Sigma Xi, Eta Kappa Nu and the Electromagnetics Academy. He was elected vice-president and president of the IEEE Antennas and Propagation Society in 1994 and 1995, respectively. He was appointed an IEEE AP-S Distinguished Lecturer and presented lectures internationally. He was elected a Fellow of IEEE in 1985 and a Fellow of Institute of Advances in Engineering (IAE) in 1986. He was also a member of the Strategic Planning and Review Committee (SPARC) of the IEEE. He was the IEEE AP-S Los Angeles Chapter Chairman (1987–1989); his chapter won the best chapter awards in two consecutive years. He has been the plenary and millennium session speaker at numerous national and international symposia. He has been the organizer and presenter of many successful short courses worldwide. He was one of the directors and vice president of the Antennas Measurement AMTA for three years. He has also served as chairman and co-chairman of several national and international symposia. He was also a member of the University of California at Los Angeles (UCLA) Graduate council for three years.

For his contributions, Dr. Rahmat-Samii has received numerous NASA and JPL Certificates of Recognition. In 1984, Prof. Rahmat-Samii was the recipient of the coveted Henry Booker Award of International Union of Radio Science (URSI), which is given triennially to the most outstanding young radio scientist in North America. Since 1987, he has been designated every three years as one of the Academy of Science's Research Council Representatives to the URSI General Assemblies held in various parts of the world. He was also invited speaker to address the URSI 75th anniversary in Belgium. In 1992 and 1995, he was the recipient of the Best Application Paper Prize Award (Wheeler Award) for papers published in 1991 and 1993 IEEE AP-S Transactions. From 1993 to 95, three of his Ph.D. students were named the Most Outstanding Ph.D. Students at the School of Engineering and Applied Science, UCLA. Ten others received various Student Paper Awards at the 1993–2004 IEEE AP-S/URSI Symposia. In 1999, he was the recipient of the University of Illinois ECE Distinguished Alumni Award. In 2000, Prof. Rahmat-Samii was the recipient of IEEE Third Millennium Medal and the AMTA Distinguished Achievement Award. In 2001, Rahmat-Samii was the recipient of the Honorary Doctorate in physics from the University of Santiago de Compostela, Spain. In 2001, he was elected as a Foreign Member of the Royal Flemish Academy of Belgium for Science and the Arts. In 2002, he received the Technical Excellence Award from JPL. He is the winner of the 2005 URSI Booker Gold Medal presented at the URSI General Assembly.

Jaehoon Kim received the B.S. degree in electronics from Kyungpook National University, Daegu, Korea, in 1993, the M.S. degree in electronic and electrical engineering from the Pohang University of Science and Technology, Pohang, Korea, in 1996, and the Ph.D. degree in electrical engineering at the University of California at Los Angeles (UCLA) in 2005.

From 1996 to 2001, he was a Research Engineer with the SK Telecom Research and Development Center, Kyunggi, Korea. In 2005, he was a Post-Doctorate Researcher at UCLA antenna lab. From 2006, Dr. Kim has been working as a R&D Manager at Fractus S.A. in Barcelona, Spain. His main research interest is RF technology for wireless communications and biomedical applications. He was the recipient of the Best Student Paper Award presented at the 2003 Antenna Measurement Techniques Association (AMTA) Symposium. He was the student paper finalist for IEEE AP-S International Symposium and USNC/URSI in 2004.

Printed in the United States
by Baker & Taylor Publisher Services